AF251812

Interference and Resource Management in Heterogeneous Wireless Networks

Interference and Resource Management in Heterogeneous Wireless Networks

Jiandong Li

Min Sheng

Xijun Wang

Hongguang Sun

ARTECH

HOUSE

BOSTON | LONDON
artechhouse.com

Library of Congress Cataloging-in-Publication Data
A catalog record for this book is available from the U.S. Library of Congress.

British Library Cataloguing in Publication Data
A catalogue record for this book is available from the British Library.

Cover design by John Gomes

ISBN 13: 978-1-63081-340-6

© 2018 ARTECH HOUSE
685 Canton Street
Norwood, MA 02062

10 9 8 7 6 5 4 3 2 1

Contents

Chapter 1

Overview of 5G Heterogeneous Wireless Networks

Providing massive device connectivity, enormous network capacity, and mega user experienced data rates are among the aggressive targets of the fifth generation (5G) wireless communications systems. In particular, as broadly foreseen for the upcoming 5G wireless communication networks, the data transmission rate desired by each user is approaching 10 Mb/s or more for high-definition (HD) video service and 100 Mb/s to support services such as ultrahigh-definition TV (UHDTV) [1], even in high mobility scenarios. In the near future other emerging applications (e.g., interactive 3D video conferences) may require even higher transmission rates up to 10 Gb/s [1]. Following this trend, it is expected that the requirement for global mobile data traffic in 2030 will be increased by 20,000-fold compared to that in 2020, and device connections will reach 100 billion [2].

To circumvent the above issues, the heterogeneous wireless network (HetNet) has been proposed as a promising candidate in the upcoming 5G systems, which facilitates the progress of new services, as well as fulfills the user quality of experience (QoE). Generally, a HetNet consists of the base stations (BSs) of various types, including macrocell BSs and small-cell BSs such as picocell BSs and femtocell BSs. Macrocell BSs are predeployed to provide broad coverage for users, while small-cell BSs like femtocell BSs of plug-and-play features may provide on-demand services and ubiquitous coverage for users. Through deploying small cells, indoor and hotspot coverage of cellular networks could be significantly improved. Moreover, the network capacity is significantly increased thanks to higher spatial spectrum reuse.

Another technique to address the explosion of mobile data traffic is device-to-device (D2D) communications [3,4]. With this technology, mobile users in prox-

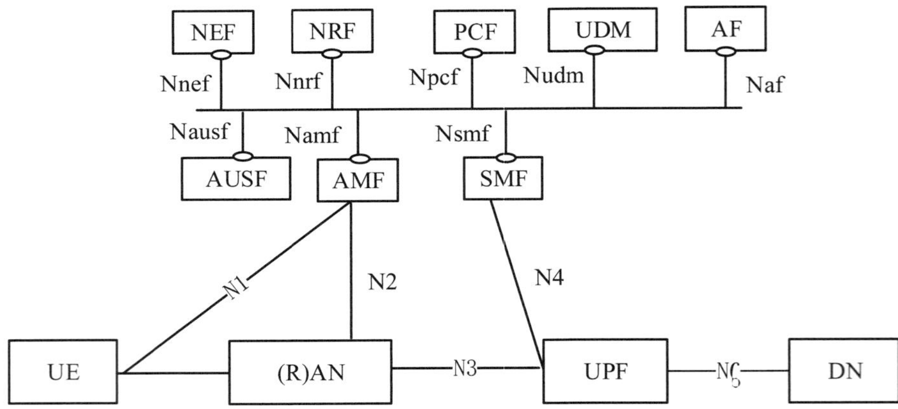

Figure 1.1: 5G system architecture.

imity can establish a direct link and bypass the base stations, thereby offloading the network infrastructure and providing increased spectral efficiency [5–7]. Considering the coexistence of cellular users and D2D users, D2D communications can be generally classified into underlaying and overlaying modes. The available spectrum resources could be universally shared by cellular and D2D users in the underlaying mode, while spectrum resources are shared by cellular and D2D users in a dedicated pattern in the overlaying mode. By comparison, spectrum resources are more likely to be fully exploited in the underlaying mode. Therefore, D2D communications underlaying cellular networks have great potential to be applied in the 5G systems.

In the following, we first provide a comprehensive discussion on the architectures of 5G HetNet. As two typical scenarios, the fundamentals of small cell and D2D communications underlaying cellular networks are then described in detail.

1.1   Architecture of 5G Heterogeneous Wireless Networks

The 5G system architecture is defined to support data connectivity and services enabling deployments to use techniques such as network function virtualization and software defined networking [8]. The 5G system architecture leverages service-based interactions between control plane network functions where identified.

Figure 1.1 depicts the non-roaming reference architecture, which consists of the following network functions (NF):

- Authentication server function (AUSF)
- Core access and mobility management function (AMF)

- Data network (DN); for example, operator services, Internet access, or third-party services
    - Structured data storage network function (SDSF)
    - Unstructured data storage network function (UDSF)
    - Network exposure function (NEF)
    - NF repository function (NRF)
    - Policy control function (PCF)
    - Session management function (SMF)
    - Unified data management (UDM)
    - User plane function (UPF)
    - Application function (AF)
    - User equipment (UE)
    - (Radio) access network ((R)AN)

We then describe the functionalities of several key network functions in the following.

Network Exposure Function. The NEF provides a means to securely expose the services and capabilities provided by 3GPP network functions. As well, NEF translates between information exchanged with the AF and information exchanged with the internal network function. Besides, it receives information from other network functions (based on exposed capabilities of other network functions).

Network Function Repository Function. The NRF supports service discovery function and maintains the information of available NF instances and their supported services.

Policy Control Function. The PCF supports a unified policy framework to govern network behavior and provides policy rules to control plane function(s) to enforce them. In addition, the PCF is capable of implementing a front end to access subscription information relevant for policy decisions in a user data repository (UDR).

Unified Data Management (UDM). The UDM includes two parts: the application front end (FE) and the UDR. As shown in Figure 1.2, the UDM reference architecture consists of UDM FE, which is in charge of processing of credentials, location management, subscription management and so on, and PCF, which is in charge of policy control.

Application Function. The AF interacts with the 3GPP core network in order to provide services. Based on operator deployment, the AF considered to be trusted by the operator can be allowed to interact directly with relevant network functions. The AF not allowed by the operator to directly access the network functions use the external exposure framework via the NEF to interact with relevant network functions.

AUSF Selection Function. The AMF performs AUSF selection to allocate an AUSF that performs authentication between the UE and 5G CN in the HPLMN. The AUSF selection function in the AMF utilizes the NRF to discover the AUSF

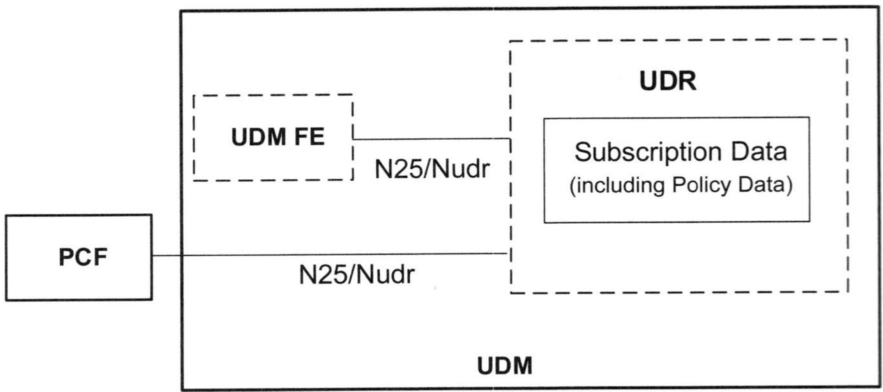

Figure 1.2: UDM reference architecture.

instance(s) unless AUSF information is available by other means (e.g. locally configured on AMF) and select an AUSF instance based on the obtained AUSF information.

AMF Selection. The AMF selection functionality is applicable to both 3GPP access and non-3GPP access. The AMF selection functionality can be supported by the 5G-AN (e.g., RAN and N3IWF) and is used to select an AMF for a given UE. An AMF supports the AMF selection functionality to select an AMF for relocation or because the initially selected AMF was not an appropriate AMF to serve the UE (e.g., due to change of allowed NSSAI based on subscription).

SMF Selection Function. The SMF selection function is supported by the AMF and is used to allocate an SMF that manages the PDU session. The SMF selection function in the AMF shall utilize the network repository function to discover the SMF instance(s) unless SMF information is available by other means (e.g., locally configured on AMF). The NRF provides the IP address or the FQDN of SMF instance(s) to the AMF.

1.2   Radio Access Networks in 5G

After giving a brief introduction to the potential 5G architecture, we primarily focus upon the RAN in 5G. In the following, we introduce several typical RAN options, including traditional cellular architecture, cloud RAN (C-RAN), fog RAN (F-RAN) and cloudlet RAN, which could serve as the promising and feasible RANs in 5G heterogeneous wireless network.

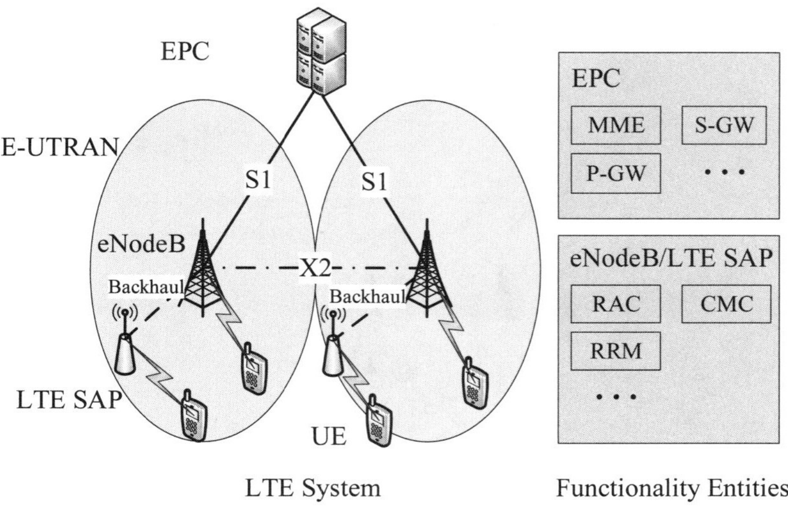

Figure 1.3: Traditional cellular architecture in LTE system.

1.2.1 Traditional Cellular Architecture in the Long-Term Evolution System

It is shown in Figure 1.3 that the LTE system is comprised of the core network (i.e., evolved packet core (EPC)) and the access network (i.e., evolved universal terrestrial radio access network (E-UTRAN)) [9]. In the EPC, functional entities such as mobility management entity (MME), serving gateway (S-GW), and packet data network gateway (P-GW) are involved. In E-UTRAN, essential elements include the eNodeB and small cell access point (SAP). To serve as a bridge between EPC and user equipment (UE), eNodeBs and LTE SAPs are responsible for all radio-related functions, including radio admission control (RAC), connection mobility control (CMC), radio resource management (RRM) and so on.

It is observed from Figure 1.3 that backhaul is needed to connect the SAPs to the eNodeBs, core network, internet, and other services. For in-building use, existing broadband internet can be used. In urban outdoors, mobile operators consider this more challenging than macrocell backhaul because (a) small cells are typically in hard-to-reach, near-street-level locations rather than in more open, above-rooftop locations and (b) carrier-grade connectivity must be provided at much lower cost per bit. In one survey, 55% of operators listed backhaul as one of their biggest challenges for small cell rollout. Many different wireless and wired technologies have been proposed as solutions, and it is agreed that a toolbox of these will be needed to address a range of deployment scenarios. An industry consensus view of how the different solution characteristics match with requirements is published by the Small Cell Forum. The backhaul solution is influenced by a number of factors, including the operator's original motivation

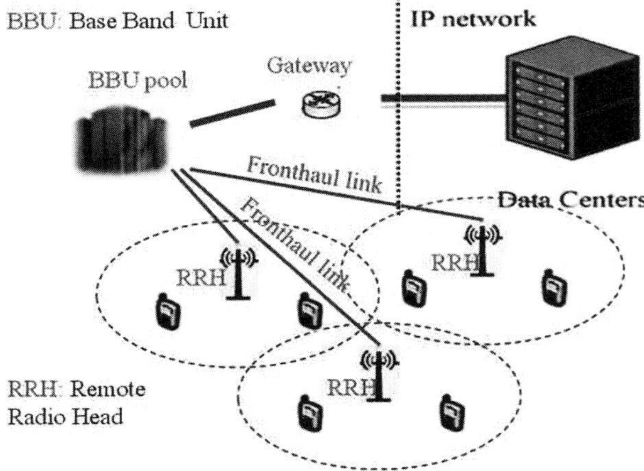

Figure 1.4: Illustration of C-RAN architecture.

to deploy small cells, which could be for targeted capacity, or indoor or outdoor coverage.

While the cellular architecture has been extensively applied, there are several limitations that hinder its application in future wireless networks. First, it is expensive to build and operate the eNodeB. The size and power of an electrical system is significantly reduced with the aid of Moore's law, but the supporting facilities of eNodeBs are not improved as well. Secondly, network densification has become a general trend in the future wireless network so as to improve user and network performance. When more network infrastructures, such as eNodeBs, are added and deployed, interference among cellular transmission is more severe as eNodeBs are closer to each other, because most are reusing the same frequency due to the limitation of available spectrum resources. As well, the traffic of each eNodeB fluctuates, which is called a tide effect, due to user mobility, and as a consequence, the utilization of an individual base transceiver station is pretty low and could be significantly improved. Aware of the above shortcomings, novel architectures have been emerging to serve as the complementary, the detail of which is discussed next.

1.2.2   Cloud RAN Architecture

C-RAN could be viewed as an architectural evolution of the above distributed cellular/LTE system. It could fully take advantage of many technological advances in wireless and wired communications systems. Specifically, for C-RAN (see Figure 1.4), its advantage lies into the fact that the functions of centralized control, storage, and signal processing are all aggregated in the cloud-based baseband processing unit pool (BBU pool) so that all the available resources could be managed in a unified pattern [10].

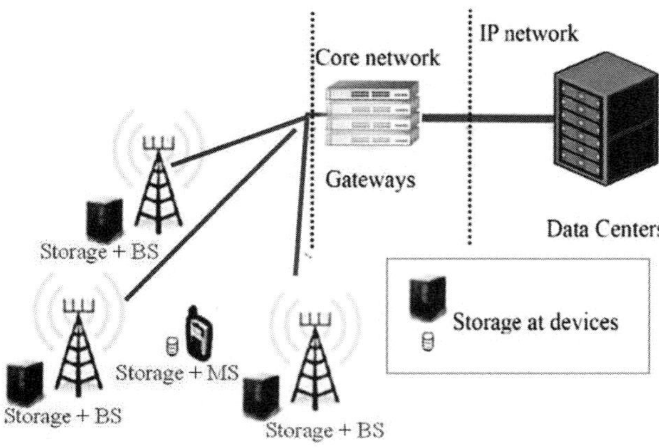

Figure 1.5: Illustration of F-RAN architecture.

More specifically, a C-RAN consists of a BBU pool and multiple remote radio heads (RRHs), where the RRHs are uniformly controlled and coordinated by the BBU pool. The BBU pool and RRHs significantly curtail capital and operating expenditure (e.g., compared to conventional cellular networks) and also support high-speed video transmissions with exceptional spectrum and energy efficiency. In particular, a powerful centralized BBU pool has natural advantages of caching and scheduling data packets and understanding the statistics of various types of traffic such as video traffic [11]. In particular, smart content caching is widely recognized as a solution for releasing the traffic burden in both wired and wireless networks. Obviously, a powerful BBU close to multiple radio access technologies (RATs) provides an advantageous location to cache content such as video content. Furthermore, by centralized coordination in a BBU, data packets could be sent to mobile users via multiple RATs at the same time, which is really beneficial for speeding up transmission rate. On the other hand, regarding the different statuses of the multiple RATs, the BBU could schedule the video packets into the matched RATs according to the required user quality of service (QoS). Thus, the C-RAN architecture could offer a chance to jointly and efficiently process, cache, and transmit various videos.

1.2.3　Fog RAN Architecture

As shown in Figure 1.5, another evolutionary network architecture F-RAN has also been recently developed, which serves as a complement of the C-RAN architecture by enabling the RRHs to be equipped with storage and signal processing functionalities [12,13]. The resulting RRHs are termed enhanced RRHs (eRRHs). In an F-RAN, edge caching is applied to prefetch the files most frequently requested to the eRRHs' local caches. Accordingly, the heavy burden of fronthaul and the BBU pool fronthaul overhead could be significantly al-

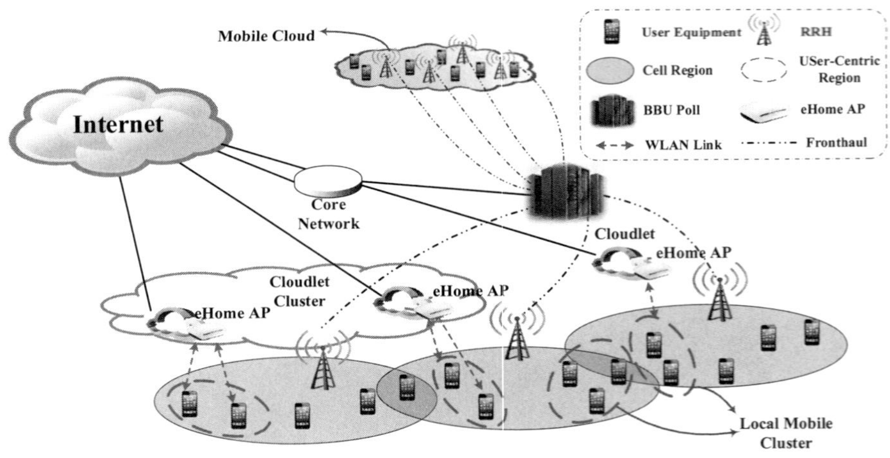

Figure 1.6: Illustration of cloudlet RAN architecture.

leviated and higher spectrum utilization could be yielded. It is worth noting that instead of minimizing the capital and operating expenditure via reduced-complexity edge nodes, the target of the F-RAN architecture is to enhance the system performance in terms of delivery rate through taking advantage of the BBU pool and caching resources [14].

1.2.4   Cloudlet RAN Architecture

Leverage the advantages of C-RAN and F-RAN, we propose an architecture, namely Cloudlet Radio Access Network (Cloudlet RAN), for the 5G mobile networks, which not only aids end users in downloading content from steadily available servers but also facilitates uploading content.

In Cloudlet RAN (see Figure 1.6), the RRH\BBU pool, fronthaul and wireless local area network (WLAN) links are as the same as their traditional definition and the biggest difference from C-RAN and F-RAN is that a new entity called a cloudlet is introduced. To be specific, a cloudlet is a trusted, resource-rich entity that is well-connected to the Internet and available for use by nearby mobile devices. The cloudlet connects users with high-bandwidth WLAN. Indeed, it would be relatively straightforward to integrate cloudlet and Wi-Fi access point hardware into a single, easily deployable entity, which is called an enhanced homeAP. According to the above discussion, a cloudlet can be referred as a private cloud.

In cloudlet RAN, multiple cloudlets can form a cloudlet cluster, which is a friend cloud and can only be accessed by friends. Different RANs and BBU pools form a mobile cloud, which can be viewed as public cloud. It is evident that the proposed architecture is a multilayer cloudlet RAN architecture, which combines the advantages of both C-RAN and F-RAN and can utilize network resources adequately.

The proposed multilayer cloudlet RAN architecture can be expressed from the following two perspectives: the physical representation and functional representation. For the physical representation, new physical entities should be added in both the wireless access part and the core network part. In the access part, homeAP is added to finish the function of cloudlet. In contrast, to facilitate the management of cloudlet and enable the formulation of Friend cloud, access controllers should be incorporated in the core network part.

For the functional representation, diverse management and control functions should be included in this architecture to fully reap the benefits promised by the cloudlet RAN. In particular, before specific management and control decisions are made, necessary contexts should be analyzed in the analysis modular placed in the related entity. Afterward, the management and control modular will operate in a different time scale and make different decisions. To be specific, for the management modular, besides traditional operation and maintenance functions, the management of cloudlets is the newly required function. The control modular is responsible for dynamic radio resource allocation, mobile handovers, resource interchange, resource migration, and so on. With the aid of these management and control functions, some promising techniques, such as self-organizing networks (SON), mobile virtual network operator (MVNO), and resource coordination can be enabled in the cloudlet RAN architecture.

1.3   5G Heterogeneous Wireless Networks

The future 5G wireless network is an integration of all types of communications systems to provide targeted QoS requirement for all users. Therefore, heterogeneous wireless network is a general trend that could encompass conventional RAN functions, RAN transport capability, small cells and Wi-Fi functionality and so on, with different operating functionalities and/or protocols.

As two typical cases, the heterogeneous networks, which integrate small cells and D2D communications into traditional macrocell systems, are discussed in detail next.

1.3.1   Small Cell Underlaying Cellular Networks

Small cells are low-powered cellular radio access nodes that operate in licensed and unlicensed spectrums that have a range of 10 meters to a few kilometers [15]. They are small compared to a mobile macrocell, partly because they have a shorter range and partly because they typically handle fewer concurrent calls or sessions.

1.3.1.1   Details of Small Cell Underlaying Cellular Networks

Capable of reusing the same frequencies many times within a geographical area, small cells could fully exploit the available spectrum. For this reason, deploying larger numbers of small cells has been recognized as a dominant method to enhance cellular network capacity and improve the quality and resilience with a growing focus using LTE Advanced [16, 17].

Small cells may encompass femtocells, picocells, and microcells. Small-cell networks can also be realized by means of distributed radio technology using centralized baseband units and remote radio heads. Beamforming technology (focusing a radio signal on a very specific area) can further enhance or focus small cell coverage. These approaches to small cells all feature central management by mobile network operators.

Small cells are available for a wide range of air interfaces including global system for mobile communication (GSM), CDMA2000, time division-synchronous code division multiple access (TD-SCDMA), wideband code division multiple access (W-CDMA), LTE and worldwide interoperability for microwave access (WiMax). In 3GPP terminology, a Home Node B (HNB) is a 3G femtocell. A Home eNodeB (HeNB) is an LTE femtocell. Wi-Fi is a small cell but does not operate in licensed spectrum and therefore cannot be managed as effectively as small cells utilizing licensed spectrum. The detail and best practice associated with the deployment of small cells varies according to use case and radio technology employed.

The most common form of small cell is a femtocell. Femtocells were initially designed for residential and small business use, with a short range and a limited number of channels. Femtocells with increased range and capacity spawned a proliferation of terms: metrocells, metro femtocells, public access femtocells, enterprise femtocells, super femtos, Class 3 femto, greater femtos and microcells. The term small cell is frequently used by analysts and industry as an umbrella to describe the different implementations of femtocells, and to clear up any confusion that femtocells are limited to residential uses. Small cells are sometimes incorrectly also used to describe distributed-antenna systems (DAS), which are not low-powered access nodes.

1.3.1.2   Challenges in Small Cell Underlaying Cellular Network from Resource Management and Interference Management Perspective

While spectrum resources could be more efficiently utilized in small cell underlaying cellular networks, which are extensively applied, critical challenges and open issues still exist that remain to be settled. Some of the challenges and potential solutions are discussed next.

Cross-tier interference. Overwhelming interference between macrocells and cochannel small cells arises from the high imbalance in path loss and transmission power between two types of cells. This may cause significant degradation in the received signal quality to mobile users. The small cells thus need to be capable of being configured, optimized, and healed so as to cause less noticeable disruption to existing networks. A self-organized networking (SON) [18, 19] is a technological enabler of such functionalities and is being introduced in Third Generation Partnership Project (3GPP) LTE-A to simplify and automate the initial provisioning, operation optimization, and maintenance of mobile networks [19, 20]. Therefore, it is a challenging problem to implement effective interference management algorithms in hyperdense small cells and heterogeneous cellular networks with SON capabilities.

Applying novel multiple access techniques in HetNets. Nonorthogonal multiple access (NOMA) has been investigated as a potential alternative to orthogonal multiuser access to further improve the spectrum efficiency of cellular networks. Therefore, the implementation of HetNets using NOMA will provide promising benefits to spectrum efficiency and energy efficiency. However, since HetNets using NOMA are designed based on the HetNets architecture, they also suffer from the severe cochannel interference and unbalanced traffic load distribution of each cell with massive deployment of small cells. Also, the non-uniform spatial distribution of mobile users will affect the performance of NOMA. Therefore, investigation of outage performance, ergodic capacity, and user fairness in NOMA schemes with spatial user distribution can be worthy work. Furthermore, it is challenging to design effective resource allocation and interference management algorithms in NOMA-enhanced HetNets.

HetNets with Cloud Computing. Considering severe intertier interference and limited cooperative gains due to constrained backhaul, as mentioned the previous sections, cloud computing technology to support global large-scale cooperative signal processing and networking can be regarded as a promising solution to improve both spectral and energy efficiency performances in HetNets. HetNets with cloud computing (HetNets-CC) [21] can be considered a cost-efficient potential solution to alleviate the overload of backhaul and complexity of cooperative processing in combination with cloud computing. As a practical HetNet-CC, C-RAN is now recognized to provide a high transmission bit rate with superior EE performance. However, the nonideal fronthaul with limited capacity [21] and long-time delay degrades performances of C-RANs, as well as how to alleviate the negative influence from the constrained fronthaul on EE performance are key challenges in HetNets-CC.

1.3.2 D2D Communications Underlaying Cellular Networks

In the coming decade, communication networks will be confronted with explosive increase in the demand for larger network capacity, more flexible network configurations, ubiquitous coverage, as well as reduction of power consumptions. D2D communications have been emerging as a feasible solution to accommodate these ongoing demands. Especially for D2D communications underlaying cellular networks, the architecture holds the promise of a series of benefits in enhancing spectrum reuse, and facilitating uniform management. Therefore, the

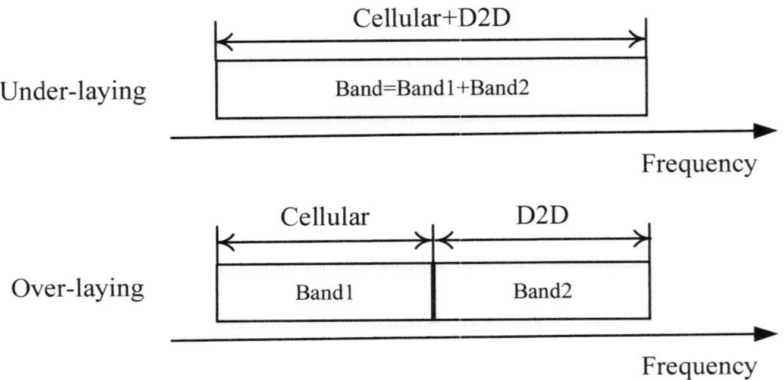

Figure 1.7: Typical coexisting modes between cellular and D2D users, i.e., underlaying and overlaying modes.

underlaying mode for D2D communications is considered as an indispensable component in LTE-Advanced systems by industries and researchers.

1.3.2.1 Details of D2D Communications Underlaying Cellular Networks

In recent years, the 3GPP has been committed to research on the proximity service (ProSe) capability in LTE [22]. D2D communications are one of the most promising proximity techniques to satisfy the wireless traffic demand. In particular, D2D communications are generally defined as technologies that enable devices in proximity to exchange information directly rather than interacting signals with the CBS. The well-known D2D technologies are Bluetooth and WiFi-Direct. In this way, proximate devices can avoid the limitation of cellular uplink and downlink. Moreover, for the network-assisted case, the utilized resources are allocated by a centralized controller, the CBS, and hence there is no need to worry about overwhelming interference. This will immensely motivate the proximate applications and services. Wherever you are, you can get connected to the neighboring devices you prefer. For instance, in your home, you can instruct the electronic devices to get all the housework done through the D2D remote control; in the subway, you can share your news or the latest hit videos with those in your surroundings using D2D to compensate for the low-quality service underground; even at a gas station, you can get a refuel by yourself with the charge paid via your D2D-enabled mobile phone. What seems impossible now will turn into realities in the future with D2D technology integrated into LTE-Advanced systems.

Considering the coexistence of cellular users and D2D users, D2D communications can be generally classified into two modes: underlaying mode [23, 24] and overlaying mode [25, 26]. As shown in Figure 1.7, the available bandwidth is universally shared by cellular and D2D users in underlaying mode. In contrast, the bandwidths, namely, Band1 and Band2 in this example, would be disjointly allocated to cellular users and D2D users. The transmissions of cellular (D2D) users will not disturb those of D2D (cellular) users in the overlaying model. Therefore, no cross-tier interference exists between cellular and D2D users. Due to the scarcity of spectrum resources, however, the available bandwidth cannot really be fully exploited by cellular and D2D users. As an example, when cellular (D2D) users are sparsely deployed and D2D (cellular) users are densely deployed, the bandwidth allocated to cellular (D2D) users is under exploited, while the bandwidth allocated to D2D (cellular) users is over exploited. The deficiency is mainly due to the coexistence of cellular and D2D in the overlaying model. For this reason, the underlaying mode is more flexible and favorable for practical concerns.

Meanwhile, the cross-tier interference from D2D users to cellular users cannot be much too overwhelming, and so the transmit power of D2D users should be strictly limited. As a consequence, it may result in higher spatial reuse and a lower interference level, which contributes to enhancing the network capacity. Furthermore, in the underlaying mode, some state-of-the-art techniques, such as network coding and cooperative communications, can be more readily applied by cellular users and D2D users aided by the centralized controller (i.e., CBS). Therefore, the underlaying mode could help alleviate the potential bottleneck and provide higher data rates for proximity services.

1.3.2.2 Challenges in D2D Underlaying Cellular Network from a Resource Management and Interference Management Perspective

Despite the above benefits, however, the underlaying mode for D2D communications may encounter several challenges as well, which are discussed as follows.

Among others, cross-tier interference exists between cellular users and D2D users when system bandwidth is universally occupied by them. In cellular networks, bandwidth is basically uniquely allocated to each cellular user within one cell to avoid intercell interference so as to guarantee the QoS of cellular users. When one or more D2D users exist to share bandwidth with the cellular users, cross-tier interference unavoidably appears. Hence, it is of great importance to handle the cross-tier interference via techniques, such as power control and resources allocation, to ensure the QoS of both cellular users and D2D users.

Another critical issue that may hinder the application of the underlaying mode is the centralized management of cellular users and D2D users. To provide a better service for cellular users and D2D users, centralized management is more preferable compared to decentralized one. However, it is estimated that the number of parameters to be configured and optimized by one network node would be over 2000 in the 5G networks [27]. Although users are with higher computation capability or even novel architectures such as C-RAN and F-RAN is adopted, it is no longer realistic to implement management of cellular and D2D users in a centralized pattern. Regarding this, management should be instead implemented in a distributed and self-organizing fashion. Meanwhile, a distinctive requirement for a well-designed self-organizing management is densability, in addition to scalability, stability and agility. In particular, densability means the overhead and complexity of the resource management and interference management solutions should not increase unboundedly with network densification. Specifically, when network performance increases with network densification, the overhead and complexity should not grow faster and, in turn, lead to performance deterioration.

According to the above discussion, the challenges lying ahead should be overcome first before the possible application of D2D communications underlaying cellular networks.

REFERENCES

[1] Datang Telecom Technology & Industry Group, "Evolution, convergence, and innovation: 5G white paper," Tech. Rep., Dec. 2013. [Online]. Available: http://www.imt-2020.org.cn/en/documents/download/3

[2] IMT-2020 (5G) Promotion Group, "5G vision and requirements," Tech. Rep., Dec. 2015. [Online]. Available: http://www.imt-2020.org.cn/en/documents/download/3

[3] D. Feng, L. Lu, Y. Yuan-Wu, G. Y. Li, S. Li, and G. Feng, "Device-to-device communications in cellular networks," *IEEE Commun. Mag.*, vol. 52, no. 4, pp. 49–55, Apr. 2014.

[4] L. Song, D. Niyato, Z. Han, and E. Hossain, "Game-theoretic resource allocation methods for device-to-device communication," *IEEE Wireless Commun.*, vol. 21, no. 3, pp. 136–144, Jun. 2014.

[5] X. Lin, J. G. Andrews, and A. Ghosh, "Spectrum sharing for device-to-device communication in cellular networks," *IEEE Trans. Wireless Commun.*, vol. 13, no. 12, pp. 6727–6740, Dec. 2014.

[6] H. ElSawy, E. Hossain, and M. S. Alouini, "Analytical modeling of mode selection and power control for underlay D2D communication in cellular networks," *IEEE Trans. Commun.*, vol. 62, no. 11, pp. 4147–4161, Nov. 2014.

[7] M. Sheng, J. Liu, X. Wang, Y. Zhang, H. Sun, and J. Li, "On transmission capacity region of D2D integrated cellular networks with interference management," *IEEE Trans. Commun.*, vol. 63, no. 4, pp. 1383–1399, Apr. 2015.

[8] 3rd Generation Partnership Project (3GPP), "System architecture for the 5G system; technical specification group services and system aspects," TS 23.501, Tech. Rep., Jun. 2017. [Online]. Available: http://www.imt-2020.org.cn/en/documents/download/3

[9] S. Stefania, I. Toufik, and M. Baker, *LTE-the UMTS long term evolution.* London, U.K.: John Wiley, 2015.

[10] M. Peng, Y. Li, Z. Zhao, and C. Wang, "System architecture and key technologies for 5G heterogeneous cloud radio access networks," *IEEE Network*, vol. 29, no. 2, pp. 6–14, Mar. 2015.

[11] X. Wang, M. Chen, T. Taleb, A. Ksentini, and V. C. M. Leung, "Cache in the air: exploiting content caching and delivery techniques for 5G systems," *IEEE Commun. Mag.*, vol. 52, no. 2, pp. 131–139, Feb. 2014.

[12] M. Peng, S. Yan, K. Zhang, and C. Wang, "Fog-computing-based radio access networks: issues and challenges," *IEEE Network*, vol. 30, no. 4, pp. 46–53, Jul. 2016.

[13] S. Bi, R. Zhang, Z. Ding, and S. Cui, "Wireless communications in the era of big data," *IEEE Commun. Mag.*, vol. 53, no. 10, pp. 190–199, Oct. 2015.

[14] M. Tao, E. Chen, H. Zhou, and W. Yu, "Content-centric sparse multicast beamforming for cache-enabled cloud RAN," *IEEE Trans. Wireless Commun.*, vol. 15, no. 9, pp. 6118–6131, Sep. 2016.

[15] T. Nakamura, S. Nagata, A. Benjebbour, Y. Kishiyama, T. Hai, S. Xiaodong, Y. Ning, and L. Nan, "Trends in small cell enhancements in LTE advanced," *IEEE Commun. Mag.*, vol. 51, no. 2, pp. 98–105, Feb. 2013.

[16] D. Lpez-Prez, M. Ding, H. Claussen, and A. H. Jafari, "Towards 1 Gbps/UE in cellular systems: Understanding ultra-dense small cell deployments," *IEEE Commun. Surveys Tutorials*, vol. 17, no. 4, pp. 2078–2101, Fourthquarter 2015.

[17] M. Ding, D. Lopez-Perez, G. Mao, P. Wang, and Z. Lin, "Will the area spectral efficiency monotonically grow as small cells go dense?" in *Proc. IEEE GLOBECOM*, San Diego, CA, Dec. 2015, pp. 1–7.

[18] O. K. Tonguz and E. Yanmaz, "The mathematical theory of dynamic load balancing in cellular networks," *IEEE Trans. Mobile Comput.*, vol. 7, no. 12, pp. 1504–1518, Dec. 2008.

[19] H. Wang, L. Ding, P. Wu, Z. Pan, N. Liu, and X. You, "Qos-aware load balancing in 3GPP long term evolution multi-cell networks," in *Proc. IEEE ICC*, Kyoto, Japan, Jun. 2011, pp. 1–5.

[20] H. Zhang, X. s. Qiu, L. m. Meng, and X. d. Zhang, "Achieving distributed load balancing in self-organizing LTE radio access network with autonomic network management," in *Proc. IEEE GLOBECOM*, Miami, Florida, Dec. 2010, pp. 454–459.

[21] W. Wang, V. K. N. Lau, and M. Peng, "Delay-aware uplink fronthaul allocation in cloud radio access networks," *IEEE Trans. Wireless Commun.*, vol. PP, no. 99, pp. 1–1, 2017.

[22] 3rd Generation Partnership Project (3GPP), "Technical specification group services and system aspects; feasibility study for proximity services (ProSe)," TR 22.803, Tech. Rep., Jun. 2013. [Online]. Available: http://www.imt-2020.org.cn/en/documents/download/3

[23] L. Lei, Z. Zhong, C. Lin, and X. Shen, "Operator controlled device-to-device communications in LTE-advanced networks," *IEEE Wireless Commun.*, vol. 19, no. 3, pp. 96–104, Jun. 2012.

[24] X. Lin, J. G. Andrews, A. Ghosh, and R. Ratasuk, "An overview of 3GPP device-to-device proximity services," *IEEE Commun. Mag.*, vol. 52, no. 4, pp. 40–48, Apr. 2014.

[25] J. Wang, D. Zhu, C. Zhao, J. C. F. Li, and M. Lei, "Resource sharing of underlaying device-to-device and uplink cellular communications," *IEEE Commun. Lett.*, vol. 17, no. 6, pp. 1148–1151, Jun. 2013.

[26] M. Zulhasnine, C. Huang, and A. Srinivasan, "Efficient resource allocation for device-to-device communication underlaying LTE network," in *Proc. IEEE WiMob*, Niagara Falls, Canada, Oct. 2010, pp. 368–375.

[27] A. Imran, A. Zoha, and A. Abu-Dayya, "Challenges in 5G: how to empower SON with big data for enabling 5G," *IEEE Network*, vol. 28, no. 6, pp. 27–33, Nov. 2014.

Chapter 2

Network Modeling and Performance Analysis of Heterogeneous Wireless Networks

2.1 Multi-tier Cellular Networks: Modeling, Analysis, and Guidelines

2.1.1 Introduction

To meet the requirement of the increasingly growing data traffic, wireless network capacity is expected to enhance 1,000-fold by 2020. The three main methods to improve network capacity are spectrum extension, spectral efficiency improvement, and network density increase [1]. Of the time, increasing network density is regarded as the most effective way by deploying low power BSs in hot spots such as micro/pico/femtocells. Meanwhile, with the aid of multihop relays at the boundaries, conventional cellular networks can both improve coverage area and increase the capacity [2,3]. Today, cellular networks become more and more complex and have evolved into the following two types: multihop cellular networks (MCNs) [2, 4], and heterogeneous cellular networks (HCNs) [5]. In MCNs, multihop relaying technologyindexmultihop relaying technology, which has been widely used in ad hoc networks, is introduced to conventional single-hop cellular network to enhance the network capacity. HCNs deploy various kinds of low-power BSs, including micro-, pico-, and femtocells, overlaid with conventional cellular networks. Consequently, multihop heterogeneous cellular networks (MHCNs) deploying various low-power base stations overlaid with a conventional cellular network are the foundation of future wireless networks where the users can associate with the BSs via a single hop or multiple hops with the help of other users acting as relay nodes.

As one of the most important system performance metrics to evaluate wireless communication systems, capacity analysis has attracted much attention for a long time. In contrast with the channel capacity, which can be derived by Shannon theorem directly, the capacity of a wireless network is more difficult to define and calculate [6]. There has been extensive research devoted to solving this problem, which falls into the following two main classes: capacity scaling laws [7–10] and transmission capacity [11, 12]. Although capacity scaling laws provide important insights into the performance of wireless networks varying with the number of nodes, they neglect the effects of other important network parameters, such as BS density, transmit power, and average signal-to-interference-plus-noise-ratio (SINR). Moreover, transmission capacity is defined as the number of successful transmissions taking place in the network per unit area subject to a constraint on outage probability [12]. However, it can only quantify the achievable rates in single-hop transmission. In this chapter, we define the capacity as *the maximum data-rate in bits per second that can be reliably transferred from transmitter to receiver* [6]. We derive the capacity of MHCNs and present the impacts of some important network parameters (e.g., BS density and transmit power), on the capacity in a theoretical sense.

The capacity of MCNs composed of m BSs and n nodes was investigated and the per-node throughput scaling factor of $\log_2 n$ compared with traditional cellular networks was obtained [3]. The influences of system parameters, such as spatial reuse factor, user density and single-hop transmit distance on the capacity of MCNs were analyzed in [13–15]. Although simulation results were used to reveal the impacts of user density and traffic volume on the capacity gain of MCNs [16], effects of the above parameters are not quantified in prior work. With tools from stochastic geometry, HCNs are modeled where the location of BSs in HCNs as K-tier independently distributed Poisson point processes (PPPs) [17, 18]. SINR is a key to analyzing the outage and throughput in HCNs and its distribution can be obtained in closed form in [18]. By calculating the complementary cumulative distribution function (CCDF) of received SINR, the coverage probability for a randomly selected user, thus the average rate, was derived [17]. If all the users are allowed to access all the BSs, one interesting conclusion is that when all the BSs have the same SINR threshold, the change in the number of tiers or their densities and their transmit powers cannot change the coverage probability in an interference-limited scenario.

In this chapter, we consider a multihop heterogeneous cellular network consisting of K-tier randomly located BSs whose locations are modeled as homogeneous PPPs. Each tier distinguishes with different BS density, transmit power, and SINR threshold. The locations of mobile users are also assumed as an independent homogeneous PPP. Under the constraint of Rayleigh fading channels, we obtain the performance of MHCNs in terms of coverage probability and capacity. We then analyze the effect of system parameters (i.e., SINR threshold, BS density, and transmit power) on the coverage probability and capacity.

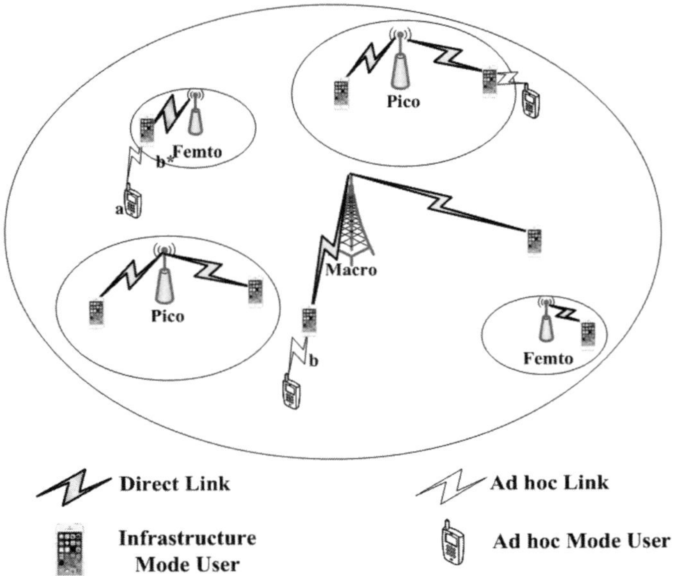

Figure 2.1: Illustration of a multihop MHCN consisting of macro-, pico-, and femtocell BSs and mobile users.

2.1.2 MHCN Model

Before describing the MHCN model, we first present the following definitions.

1. *Capacity:* The capacity of a communication system is the maximum data-rate in bits per second that can be reliably transferred from transmitter to receiver [6]. The capacity is an unsurpassable upper bound that can only be approached in practice.

2. *Coverage Probability*: The coverage probability is the probability that a typical mobile user successfully connects to BSs via single hop or multiple hops only with the constraint of its SINR requirement.

3. *Infrastructure Mode*: We say a typical user connects to a certain BS in tier i $(1 \leq i \leq K)$ in infrastructure mode only if its SINR is larger than the corresponding threshold β_i.

4. *Ad Hoc Mode*: A typical user is said to be associated with BSs in ad hoc mode if it can only connect to BSs via multiple hops. In this chapter, we only consider the two-hop access scenario where the users are in ad hoc mode. That means the users in ad hoc mode will associate with the users who can connect to BS directly and try to access BSs in two hops.

5. *Direct Link*: Direct link is the link between a BS and a user.

6. *Ad hoc Link*: Ad hoc link is the link between two users.

Figure 2.1 illustrates a MHCN with three-tier BSs (macro-, pico-, and femtocell BSs) and mobile users. Although only a single macrocell is shown for the sake of simplicity, we consider the scenario of multiple macrocells in our analysis. As in [17], we model different classes of BSs as K-tier (here $K = 3$) randomly located network nodes where each tier represents a particular type of BSs. They are distinguished by corresponding BS density, transmit power, and data rate. For example, macrocells have lower BS density and higher transmit power than those of pico- and femtocells. The BSs in the ith tier are spatially distributed as an independent homogeneous PPP ϕ_i of density λ_i, transmit at power P_i and have a particular SINR target β_i. The mobile users are also modeled as an independent homogeneous PPP ϕ_u of density λ_u, transmit at power P_u, and have a particular SINR target β_u.

The total bandwidth B is divided into two parts: θB for the transmission on direct link, and $(1 - \theta) B$ for the transmission on ad hoc link. Note that $0 < \theta < 1$. Since the transmission on direct link and ad hoc link work at different frequency bands, there is no interference between them. Therefore, the interference of the mobile users in infrastructure mode comes from all the concurrent transmitting BSs, while the interference of the mobile users in ad hoc mode comes from all the transmission nodes in the same mode. We assume that the mobile users have two network interfaces so that they can work on these two different frequency bands without any mutual interference. In other words, the mobile users who act as relays can receive their own data from BSs and transmit the relayed traffic to the users in ad hoc mode at the same time. As internet data traffic has become dominant in cellular networks, the modeling within this chapter focuses on the downlink that transfers the majority of the data [3]. Therefore, the transmission in infrastructure mode can only be transmitted over the direct link and the transmission in ad hoc mode is transmitted first over the direct link and then over the ad hoc link.

In this chapter, we only focus on open access strategy [17] with SINR being the only constraint for user association. More precisely, a mobile user can successfully associate with a BS in the ith tier via infrastructure mode only if its SINR with respect to that BS is greater than the corresponding threshold β_i. If the mobile user cannot directly associate with any BS, it will attempt to associate with the nearest user who is capable of communicating with BSs directly and connect to the BSs in ad hoc mode. We assume fading between any pair of nodes is independent Rayleigh fading and the power fading coefficient between a node x and a node y is denoted by $\gamma(x, y) \sim exp(1)$. The path loss law is the standard power law $l(x) = \|x\|^{-\alpha}$, where $\alpha > 2$ is the path loss exponent in outdoor environments and $\| x \|$ denotes the distance between the transmitter and its receiver. Thus, the received power of a typical node a from any node x_i (a and x_i also denote the position of these two nodes, respectively) can be expressed as $P_{x_i}\gamma(a, x_i)\|x_i - a\|^{-\alpha}$, where P_{x_i} is the transmission power of node x_i. We consider the Shannon channel capacity that is calculated as $R(a, x_i) = B(a, x_i) \log\left(1 + SINR(a, x_i)\right)$, where $B(a, x_i)$ is channel bandwidth and $SINR(a, x_i)$ is the SINR received by node a from node x_i. We can obtain that

$$SINR(a, x_i) = \frac{P_{x_i}\gamma(a, x_i)\|x_i - a\|^{-\alpha}}{\sum_{x\in\varphi, x\neq x_i} P_x\gamma(a, x)\|x - a\|^{-\alpha} + \sigma^2}, \tag{2.1}$$

where φ denotes the interference set of node a and σ^2 is the constant additive noise power. In the same cell, we can use orthogonal channels such as OFDM or nonoverlapping subchannels to avoid the intracell interference. Therefore, the intracell interference is not included in our analysis. For simplicity, we only consider an interference-limited scenario where the noise power is neglected.

2.1.3 Coverage Probability of MHCN

To derive the capacity of MHCN, we first investigate the coverage probability of MHCN. A typical user is said to be in coverage if and only if it connects to BSs exclusively either in infrastructure mode or in ad hoc mode if it cannot operate in infrastructure mode. If the SINR between a user and a BS is greater than the corresponding threshold, the user will operate in infrastructure mode. Otherwise, it will try to operate in ad hoc mode. We denote P_{c_i} and P_{c_a} as the coverage probabilities of a typical user associated with BSs in infrastructure mode and ad hoc mode, respectively. Since these two access modes are mutually exclusive and cannot happen together, the coverage probability of MHCNs P_{c_mh} can be expressed as $P_{c\_mh} = P_{c\_i} + P_{c\_a}$. To derive the coverage probability of MHCNs, we first calculate the coverage probability of a typical user in these two modes and then study the impacts of BS density, transmit power, and SINR threshold on coverage probability.

2.1.3.1 Coverage Probability of Users in Infrastructure Mode

If the received SINR of a typical mobile user is greater than a certain threshold, it will connect to the corresponding BS in infrastructure mode. Owing to the spatial stationarity of PPP, all the mobile users have the same statistics of received signal [19]. Therefore, we can conduct the analysis on a typical user a located at the origin. The coverage probability of users in infrastructure mode is expressed as

$$P_{c\_i} = P\left\{ \bigcup_{i \in K,\, x_i \in \phi_i} SINR(a, x_i) > \beta_i \right\}, \tag{2.2}$$

where ϕ_i denotes the set of the ith tier BSs and β_i is the corresponding SINR threshold. Under the assumption that $\beta_i > 1$, $\forall i$, the mobile user can connect to at most one BS at any time [17]. Thus, the coverage probability is the sum of the probability that the mobile user connects to each BS.

Under open access strategy, the mobile user can connect to any BS directly with its received SINR above the threshold without any extra restriction. Therefore, the probability that a mobile user connects to a typical BS is the CCDF of the received SINR. It can be calculated by determining the Laplace transform of corresponding cumulative interference I_{x_i} of the mobile user in infrastructure mode. Note that the intracell interference can be canceled by using the orthogonal resources. Since there is no interference between the direct link and ad hoc link, I_{x_i} is expressed as $I_{x_i} = \sum_{j=1}^{K} \sum_{x \in \phi_j,\, x \neq x_i} P_j \gamma(a, x) \|x - a\|^{-\alpha}$. The corresponding SINR can be denoted by

$$SINR(a, x_i) = \frac{P_i \gamma(a, x_i) \|x_i - a\|^{-\alpha}}{\sum_{j=1}^{K} \sum_{x \in \phi_j, x \neq x_i} P_j \gamma(a, x) \|x - a\|^{-\alpha}}. \tag{2.3}$$

The Laplace transform of I_{x_i} [17] is

$$\mathcal{L}_{I_{x_i}}(s) = exp\left(-s^{2/\alpha} C(\alpha) \sum_{j=1}^{K} \lambda_j \left(P_j\right)^{2/\alpha}\right), \tag{2.4}$$

where $C(\alpha) = 2\pi^2 csc\left(\frac{2\pi}{\alpha}\right)$. Summing up with the probability that the user associates with each BS, the coverage probability of users in infrastructure mode leads to a simple closed-form expression under the Rayleigh fading channel, given by

$$P_{c\_i} = \frac{\pi}{C(\alpha)} \frac{\sum_{i=1}^{K} \lambda_i P_i^{2/\alpha} \beta_i^{-2/\alpha}}{\sum_{i=1}^{K} \lambda_i P_i^{2/\alpha}}, \qquad \beta_i > 1. \tag{2.5}$$

2.1.3.2   Coverage Probability of Users in Ad Hoc Mode

If a typical mobile user a cannot access the BSs in infrastructure mode, it will try to connect to its nearest neighbor user b^* who is able to communicate with BSs directly. Thus, the coverage probability of a typical user a in ad hoc mode is given by

$$\begin{aligned} P_{c\_a} &= P\{O_i(a)\} \cdot P\{C_i(b^*), SINR(a, b^*) > \beta_u | O_i(a)\} \\ &= P\{O_i(a)\} \cdot P_{f\_relay}, \end{aligned} \tag{2.6}$$

where $C_i(b^*) = \underset{i \in K, \, x_i \in \phi_i}{\cup} SINR(b^*, x_i) > \beta_i$ is the event that user b^*, which is the nearest neighbor to user a, can associate with BSs directly, $O_i(a) = \underset{i \in K, \, x_i \in \phi_i}{\cap} SINR(a, x_i) < \beta_i$ represents the event that user a cannot connect to any BS in infrastructure mode, β_u denotes the SINR target of mobile users, and P_{f_relay} is the probability that the mobile user communicates with a relay user (its nearest neighbor user that has an infrastructure connection) successfully, on the condition that on the mobile user itself has no infrastructure connection.

The complement of the probability that a typical user connects to BSs in infrastructure mode is $P\{O_i(a)\} = 1 - P\{C_i(a)\} = 1 - P_{c\_i}$. To derive the coverage probability of a typical user associated with BSs in ad hoc mode, we should get P_{f_relay}. Next, we will first calculate the density of the concurrent transmitters in ad hoc mode λ_{act} conditioned on some users cannot operate in infrastructure mode. Then, we will get the interference I_{b^*} of a typical user in ad hoc mode according to the value of λ_{act} directly. Finally, P_{f_relay} will be obtained by deriving the Laplace transform of I_{b^*}.

Now, we will calculate the density of the concurrent transmitters in ad hoc mode λ_{act}. To differentiate the relay users from the mobile users, the set of relay users are called potential transmitters in ad hoc mode. The other mobile users, which cannot connect to BSs in single hop and try to associate with the BSs in multiple hops, are called the potential receivers in ad hoc mode. Only the users who can connect to the BSs in infrastructure mode can be viewed as potential transmitters in ad hoc mode. We assume that each mobile user is served as a potential transmitter independently of each other. Note that this assumption is true only if the locations of BSs are given. However, we will show later, this independence assumption leads to a good approximation by comparing our analytical results with the simulated ones. According to independent thinning [20], the resulting potential transmitters and receivers are also PPPs with densities of $\lambda_{Pt} = P_{c\_i} \cdot \lambda_u$ and $\lambda_{Pr} = \left(1 - P_{c\_i}\right) \cdot \lambda_u$, respectively.

A potential receiver will try to associate with BSs by connecting to the nearest potential transmitters. For example, in Figure 2.1, user a cannot access any BS in infrastructure mode. As a potential receiver, user a will search for potential transmitters (user b and user b^*) within its single-hop distance and select the nearest user (user b^*) to access. If two or more receivers select the same transmitter to connect, we will randomly select one of the receivers to pair up with the transmitter. Thus, the number of active links is decided by the minimum of active receivers and active transmitters.

In the case of $\lambda_{Pt} > \lambda_{Pr}$, the number of active links equals to the number of active receivers. Based on the strategy of selecting the nearest neighbor and the properties of PPPs, λ_{act} can be approximately calculated as [21]:

$$\lambda_{act} \approx \lambda_{Pt} \left(1 - \exp\left(-\frac{\lambda_{Pr}}{\lambda_{Pt}}\left(1 - \exp\left(-\pi d^2 \lambda_{Pt}\right)\right)\right)\right), \qquad (2.7)$$

where d is defined as the maximum single-hop distance in the absence of interference in ad hoc mode [21].

Since there is no interference between direct links and ad hoc links, the interference of a typical receiver in ad hoc mode is expressed as

$$I_{b^*} = \sum_{b \in \phi_{act}, b \neq b^*} P_u \gamma(a,b) \|b - a\|^{-\alpha},$$

where ϕ_{act} denotes the interference set of node a. By determining the Laplace transform of I_{b^*}, we obtain the probability that potential receiver a connects to the relay user b^* successfully conditioned on the distance d^* between these two users:

$$P\left\{C_i(b^*),\ SINR(a,b^*) > \beta_u | O_i(a), d^*\right\} = \exp\left(-C\left(\alpha\right)\beta_u^{2/\alpha} {d^*}^2 \lambda_{act}\right), \qquad (2.8)$$

where $C(\alpha) = 2\pi^2 csc\left(\frac{2\pi}{\alpha}\right)$. According to the properties of PPPs [22], the probability density function (pdf) of d^* is [23]

$$f\left(d^*\right) = 2\pi d^* \lambda_{Pt} \frac{\exp\left(-\pi {d^*}^2 \lambda_{Pt}\right)}{1 - \exp\left(-\pi d^2 \lambda_{Pt}\right)}. \qquad (2.9)$$

The probability that a typical mobile user can successfully connect to the relay user is given by

$$P_{f\_relay} = P\left\{C_i(b^*),\, SINR(a,b^*) > \beta_u | O_i(a)\right\}$$

$$= \int P\left\{C_i(b^*),\, SINR(a,b^*) > \beta_u | O_i(a), d^*\right\} f(d^*)\, \mathrm{d}d^*, \qquad (2.10)$$

Consequently, we derive [23]

$$P_{f\_relay} = \frac{\pi \lambda_{Pt}}{\left(C(\alpha)\beta_u^{2/\alpha}\lambda_{act} + \pi\lambda_{Pt}\right)} \cdot \frac{1 - \exp\left(-\left(C(\alpha)\beta_u^{2/\alpha}\lambda_{act} + \pi\lambda_{Pt}\right)d^2\right)}{1 - \exp\left(-\pi d^2\lambda_{Pt}\right)}.$$

$$(2.11)$$

Combining the above equations, the probability that a typical mobile user connects to BSs in ad hoc mode $(\lambda_{Pt} > \lambda_{Pr})$ can be expressed as [23]

$$P_{c\_a} = \frac{\left(1 - P_{c\_inf}\right)\cdot\pi}{\left(C(\alpha)\beta_u^{2/\alpha}\lambda_{act}/\lambda_{Pt} + \pi\right)} \cdot \frac{1 - \exp\left(-\left(C(\alpha)\beta_u^{2/\alpha}\lambda_{act} + \pi\lambda_{Pt}\right)d^2\right)}{1 - \exp\left(-\pi d^2\lambda_{Pt}\right)}.$$

$$(2.12)$$

Let us consider the asymptotic condition $\exp\left(-\pi d^2\lambda_{Pt}\right) \to 0$; we have

$$P_{c\_a} \approx \frac{\left(1 - P_{c\_inf}\right)\cdot\pi}{C(\alpha)\beta_u^{2/\alpha}\left(1 - \exp\left(-\frac{\left(1 - P_{c\_inf}\right)}{P_{c\_inf}}\right)\right) + \pi}. \qquad (2.13)$$

On the condition that $\lambda_{Pt} < \lambda_{Pr}$, the number of active links equals to the number of active transmitters. We get a similar result as follows:

$$P_{c\_a} \approx \frac{P_{c\_i}\cdot\pi}{C(\alpha)\beta_u^{2/\alpha}\left(1 - \exp\left(-\frac{P_{c\_i}}{\left(1 - P_{c\_i}\right)}\right)\right) + \pi}, \quad \lambda_{Pt} < \lambda_{Pr} \qquad (2.14)$$

Summarizing (2.13) and (2.14), we derive the coverage probability in ad hoc mode as

$$P_{c\_a} \approx \frac{\pi\cdot P_{min}}{C(\alpha)\beta_u^{2/\alpha}\left(1 - \exp\left(-\frac{P_{min}}{P_{max}}\right)\right) + \pi}, \qquad (2.15)$$

where we have $P_{min} = \min\left\{P_{c\_i},\, 1 - P_{c\_i}\right\}$, and $P_{max} = \max\left\{P_{c\_i},\, 1 - P_{c\_i}\right\}$.

From (2.15), we observe that the coverage probability in ad hoc mode decreases with the SINR threshold of users. Keeping the SINR threshold of users unchanged, the coverage probability of users in ad hoc mode is only dependent on the coverage probability of users in infrastructure mode.

Denoting $x = \frac{|1-2P_{c\_i}|}{2}$ $(0 \leq x \leq 0.5)$, the coverage probability in ad hoc mode is expressed as

$$P_{c\_a}(x) = \frac{(0.5 - x)\,\pi}{C\,(\alpha)\,\beta_u^{2/\alpha}\left(1 - \exp\left(-\frac{0.5-x}{0.5+x}\right)\right) + \pi}. \tag{2.16}$$

By taking the derivative with respect to x, we find the following conclusion: given SINR threshold of users, the coverage probability of users in ad hoc mode has maximum value at $P_{c\_i} = 0.5$ under the condition that $C\,(\alpha)\,\beta_u^{2/\alpha} < \frac{\pi e}{3-e}$. It means that the coverage probability of users in ad hoc mode can get maximum value only on the condition that half of the mobile users connect to the BSs in infrastructure mode. The intuition behind the result is that the number of successful connections in ad hoc mode is determined by the number of potential receivers pairing up with potential transmitters successfully. The fewer the potential transmitters or receivers, the less the mobile users can associate with BSs in ad hoc mode. In this chapter, the sum of potential transmitters and receivers is decided by the density of users which remains constant. Therefore, only when the number of potential transmitters is equal to that of potential receivers, can the coverage probability of users in ad hoc mode have the maximum value.

With coverage probability of users in infrastructure mode and ad hoc mode, the expression of the coverage probability of MHCNs can be derived as [23]

$$P_{c\_mh} = \frac{\pi}{C(\alpha)} \frac{\sum_{i=1}^{K} \lambda_i P_i^{2/\alpha} \beta_i^{-2/\alpha}}{\sum_{i=1}^{K} \lambda_i P_i^{2/\alpha}} + \frac{\pi \cdot P_{min}}{C\,(\alpha)\,\beta_u^{2/\alpha}\left(1 - \exp\left(-\frac{P_{min}}{P_{max}}\right)\right) + \pi}, \tag{2.17}$$

where we define $C(\alpha) = 2\pi^2 csc\left(\frac{2\pi}{\alpha}\right)$.

According to (2.17), the coverage probability of MHCNs is affected by the BS density, transmit power, and SINR threshold. We first derive the condition when the coverage probability of MHCNs can be improved.

Given that the SINR thresholds of various tiers are not all the same (i.e., $\beta_i \neq \beta_j$, $\exists i \neq j$, $i, j \in [1, K]$), the coverage probability of MHCNs can be improved by increasing the BS density λ_m or transmit power P_m of the mth tier on the condition that [23]

$$\beta_m^{-2/\alpha} > \frac{\sum_{i=1,i\neq m}^{K} \lambda_i P_i^{2/\alpha} \beta_i^{-2/\alpha}}{\sum_{i=1,i\neq m}^{K} \lambda_i P_i^{2/\alpha}}. \tag{2.18}$$

We call the condition in (2.18) a SINR enhancement condition. When the SINR enhancement condition is satisfied, the increase in BS density or transmit power of a certain tier will improve the SINR because the resulting increment of signal power is higher than that of interference power. Therefore, the coverage probability of MHCNs can be improved by increasing the BS density or transmit power. If the SINR enhancement cannot hold, the improvement of the BS density and transmit power of a certain tier will decrease the coverage probability of MHCNs due to intercell interference.

When all the BSs have the same SINR threshold, the coverage probability in infrastructure mode can be simplified to $P_{c\_i} = \frac{\pi}{C(\alpha)\beta^{2/\alpha}}$, which is independent of the number of tiers or their relative densities, transmit powers, and the density of users. Moreover, the coverage probability in ad hoc mode is only determined by P_{c_i}. Therefore, the coverage probability of MHCNs summing up with the coverage probabilities in infrastructure mode and ad hoc mode doesn't change with the above parameters. We observe that when all the tiers have the same SINR thresholds, the received SINR distribution does not change with the number of tiers, their relative densities, transmit powers and the density of users, and neither does the coverage probability of MHCNs. This is because a change of the above parameters leads to a change in signal power and interference power with the same factor, and thus the effects are canceled.

2.1.4   Capacity of MHCNs

In this section, we first calculate the average throughput of users in infrastructure mode and ad hoc mode separately. Then the capacity of MHCNs is derived by maximizing the average throughput of each user. Finally, we investigate the effects of BS density, transmit power, and SINR threshold on the capacity of MHCNs.

A mobile user receives data from BSs exclusively either in infrastructure mode or in ad hoc mode. Only on the condition that it cannot operate in infrastructure mode, will it access BSs in ad hoc mode. Otherwise, it will operate in infrastructure mode. Hence the average throughput of each user is expressed as

$$\overline{T}_{h\_mh} = P_{c\_i} \cdot \overline{T}_{h\_i} + P_{c\_a} \cdot \overline{T}_{h\_a}, \tag{2.19}$$

where P_{c_i} and P_{c_a} are the probabilities that the users connect to the BSs in infrastructure mode and in ad hoc mode. $\overline{T}_{h\_i}$ and $\overline{T}_{h\_a}$ are the average throughputs in these two access modes, respectively.

2.1.4.1   Average Throughput of Users in Infrastructure Mode

In infrastructure mode, data is transmitted from the BSs to the users directly. Since all the traffic is generated from BSs, the mobile users in coverage share the direct link resource θB ($0 < \theta < 1$). Under the assumption that each mobile user is allocated the same bandwidth resource or scheduled with the same probability without any difference, we have [23]

$$\overline{T}_{h\_i} = \frac{\sum_{i=1}^{K} \lambda_i \cdot \theta B}{P_{c\_mh} \cdot \lambda_u} \cdot R_{e\_i}, \tag{2.20}$$

where R_{e_i} denotes the spectral efficiency (average transmission rate in unit bandwidth) of direct links.

In this section, we also conduct the analysis on a typical user a located at the origin. The calculation of R_{e_i} is based on the conditional CCDF of SINR and will be omitted here due to space constraints. We can obtain R_{e_i} directly [17]

$$R_{e\_i} = \log\left(1 + \beta_{min}\right) + \frac{\sum_{i=1}^{K} \lambda_i P_i^{2/\alpha} A\left(\alpha, \beta_i, \beta_{min}\right)}{\sum_{i=1}^{K} \lambda_i P_i^{2/\alpha} \beta_i^{-2/\alpha}}, \tag{2.21}$$

where $\beta_{min} = \min\left\{\beta_1, \beta_2, \cdots, \beta_K\right\}$ and $A\left(\alpha, \beta_i, \beta_{min}\right) = \int_{\beta_{min}}^{\infty} \frac{\max(\beta_i, x)^{-2/\alpha}}{1+x} dx$. Inserting (2.21) into (2.20), the average throughput in infrastructure mode can be easily calculated based on a single integral. As expected, $\overline{T}_{h\_inf}$ increases with the allocated resource θB and the density of BSs, and decreases with the density of the mobile users.

2.1.4.2   Average Throughput of Users in Ad Hoc Mode

In ad hoc mode, the data generated from BSs should be first transmitted to the selected relay user and then to destination user. Since the traffic goes through two consecutive links, direct link and ad hoc link, the link with the minimum data rate becomes the bottleneck link and determines the end-to-end throughput. Thus, we obtain $\overline{T}_{h\_a} = \min\left\{\overline{R}_{direct}, \overline{R}_{adhoc}\right\}$, where $\overline{R}_{direct}$ and $\overline{R}_{adhoc}$ denote the average data rate of direct link and that of ad hoc link, respectively. The average data rate of direct link $\overline{R}_{direct}$ in ad hoc mode equals the average throughput $\overline{T}_{h\_i}$ in infrastructure mode. To get $\overline{T}_{h\_a}$, we only need to calculate $\overline{R}_{adhoc}$.

Since the concurrent transmissions improve the spatial reuse, the average data rate of the ad hoc link is denoted as [23]

$$\overline{R}_{adhoc} = \frac{\lambda_{act} \cdot \left(1 - \theta\right) B}{P_{c\_a} \cdot \lambda_u} \cdot R_{e\_a}, \tag{2.22}$$

where we have

$$R_{e\_a} = \log\left(1 + \beta_u\right) + \int_{\beta_u}^{\infty} \frac{C\left(\alpha\right) \beta_u^{2/\alpha} \lambda_{act} + \pi \lambda_{max}}{\left(1 + y\right)\left(C\left(\alpha\right) y^{2/\alpha} \lambda_{act} + \pi \lambda_{max}\right)} dy. \tag{2.23}$$

where b^* is the nearest relay user to the user a and $C_a\left(b^*\right)$ denotes the event that the relay user communicates successfully with the destination user.

When $\beta_i = \beta_u = \beta$, $\forall i$, the spectral efficiency of direct link and ad hoc link can be simplified as

$$R_{e\_i} = \log\left(1 + \beta\right) + \beta^{2/\alpha} \int_{\beta}^{\infty} \frac{1}{x^{-2/\alpha}\left(1 + x\right)} dx, \tag{2.24}$$

$$R_{e\_a} = \log\left(1 + \beta\right) + \int_{\beta}^{\infty} \frac{C\left(\alpha\right) \beta^{2/\alpha} \frac{\lambda_{act}}{\lambda_{min}} + \pi}{\left(1 + y\right)\left(C\left(\alpha\right) y^{2/\alpha} \frac{\lambda_{act}}{\lambda_{min}} + \pi\right)} dy. \tag{2.25}$$

By comparing the above two equations, we have $R_{e\_adhoc} > R_{e\_i}$, which means that when all the tiers have the same SINR threshold with that of mobile users ($\beta_i = \beta_u = \beta$, $\forall i$), the spectral efficiency of the ad hoc link is greater than that of the direct link.

2.1.4.3   The Capacity of MHCNs

According to the relationship between the data rate of direct link and ad hoc link, there are two possible situations of average throughput in MHCNs as follows,

1. When $\overline{R}_{direct} \leq \overline{R}_{adhoc}$, the average throughput of MHCNs can be simplified as $\overline{T}_{h\_mh} = \frac{\theta \cdot \sum_{i=1}^{K} \lambda_i \cdot B}{\lambda_u} \cdot R_{e\_i}$.

2. When $\overline{R}_{direct} > \overline{R}_{adhoc}$, $\overline{T}_{h\_mh}$ is denoted by $\overline{T}_{h\_mh} < \frac{\theta \cdot \sum_{i=1}^{K} \lambda_i \cdot B}{\lambda_u} \cdot R_{e\_i}$.

By combinning these two formulas, we have $\overline{T}_{h\_mh} \leq \frac{\theta \cdot \sum_{i=1}^{K} \lambda_i \cdot B}{\lambda_u} \cdot R_{e\_i}$. Thus, given the BS density, transmit power, and SINR threshold of each tier, the capacity of MHCNs is

$$C = \frac{\theta \cdot \sum_{i=1}^{K} \lambda_i \cdot B}{\lambda_u} \cdot R_{e\_i}, \tag{2.26}$$

where $R_{e\_i} = \log\left(1 + \beta_{min}\right) + \frac{\sum_{i=1}^{K} \lambda_i P_i^{2/\alpha} A(\alpha, \beta_i, \beta_{min})}{\sum_{i=1}^{K} \lambda_i P_i^{2/\alpha} \beta_i^{-2/\alpha}}$.

We observe that although the capacity of MHCNs can be improved by cell splitting, the improvement may be diminished by the division of resources. By analyzing (2.26), we derive the condition when capacity of MHCNs can be enhanced. Specifically, given that the SINR thresholds of various tiers are not all the same (i.e., $\beta_i \neq \beta_j \exists i \neq j$, $i, j \in [1, K]$), the increase in the BS density λ_m and transmit power P_m of tier m ($1 \leq m \leq K$) improves the spectral efficiency, thereby enhancing the capacity of MHCNs on the condition that [23]

$$\frac{A\left(\alpha, \beta_m, \beta_{min}\right)}{\beta_m^{-2/\alpha}} > \frac{\sum_{i=1, i \neq m}^{K} \lambda_i P_i^{2/\alpha} A\left(\alpha, \beta_i, \beta_{min}\right)}{\sum_{i=1, i \neq m}^{K} \lambda_i P_i^{2/\alpha} \beta_i^{-2/\alpha}}. \tag{2.27}$$

We call the condition in (2.27) the spectral efficiency (SE) enhancement condition, on which the corresponding spectral efficiency is improved with an increase in the BS density and transmit power. Under the SE enhancement condition, the capacity of MHCNs are improved by deploying more BSs and increasing their transmit powers. Moreover, the augment of the BSs density λ_m also increases spatial reuse. Therefore, the increment of λ_m has a better effect on improving the capacity of MHCNs.

Let's consider the special case when $\beta_i = \beta$, $\forall i$. The spectral efficiency can be expressed as

$$R_{e\_i} = log\left(1 + \beta\right) + \beta^{2/\alpha} \int_{\beta}^{\infty} \frac{x^{-2/\alpha}}{1 + x} dx. \tag{2.28}$$

then we derive the capacity of MHCNs as [23]

$$C = \frac{\theta \cdot \sum_{i=1}^{K} \lambda_i \cdot B}{\lambda_u} \cdot \left(log\left(1 + \beta\right) + \beta^{2/\alpha} \int_{\beta}^{\infty} \frac{x^{-2/\alpha}}{1 + x} dx\right). \tag{2.29}$$

From (2.29), we get that the increase of the number of BSs results in linear improvement of the network capacity due to the spectral reuse enhancement. By taking the derivative with respect to β, we derive that the capacity of MHCNs increases with SINR threshold β.

It shows that when all the tiers have the same SINR threshold, the capacity increases with SINR threshold and linearly with BS density. From (2.29), we can also obtain that the capacity decreases with the decreasing of θ since part of the resources is allocated to ad hoc links and the resources derived by each user is decreased. However, without multihop relaying technology, users with low SINR cannot access any BS and hence are incapable of receiving data from any BS. With multihop relaying technology, SINR can be improved since the path loss is reduced. As such, the users with low SINR can access BSs in ad hoc mode and receive data from the associated BS. Therefore, we conclude that multihop relaying technology can improve the capacity and complement the effects of resource partition.

2.1.4.4 Special Cases of Interest

In this section, we will utilize the capacity of MHCNs to study some special cases of interest by setting particular values for some system parameters.

1. *Heterogeneous Cellular Networks.* When $\theta = 1$, all the resources are allocated to the transmissions on direct links. As a result, mobile users cannot access BSs in ad hoc mode due to the scarcity of resources. The scenario of MHCNs is simplified to HCNs (without the multihop access mode). The capacity of HCNs is given by $C = \frac{\sum_{i=1}^{K} \lambda_i \cdot B}{\lambda_u} \cdot R_{e\_i}$, where $R_{e\_i} = \log\left(1 + \beta_{min}\right) + \frac{\sum_{i=1}^{K} \lambda_i P_i^{2/\alpha} A(\alpha,\beta_i,\beta_{min})}{\sum_{i=1}^{K} \lambda_i P_i^{2/\alpha} \beta_i^{-2/\alpha}}$. Given SINR thresholds, MHCNs have lower capacity compared with HCNs since the resources are partitioned to the transmissions on an ad hoc link. However, the multihop transmissions have the ability to reduce the single-hop distance, thus improving the average SINR experienced by the mobile users, and thereby compensating the loss of resource partition. Although we do not analyze the impacts of multihop relay technologies on the improvement of capacity in this chapter, it will be observed in our numerical results.

2. *Multihop Cellular Networks.* Assuming that $K = 1$, only the macrocell is deployed. The scenario of MHCNs is simplified to MCNs (no low-power BSs) and the corresponding capacity is derived as follows $C = \frac{\theta \lambda_1 B}{\lambda_u} \cdot R_{e\_i}$, where $R_{e\_i} = \log\left(1 + \beta_{min}\right) + \frac{\sum_{i=1}^{K} \lambda_i P_i^{2/\alpha} A(\alpha,\beta_i,\beta_{min})}{\sum_{i=1}^{K} \lambda_i P_i^{2/\alpha} \beta_i^{-2/\alpha}}$. Given θ and SINR thresholds, we can get the corresponding capacity. If the preconditions are relaxed, the average throughput may achieve its maximum value when $\theta \to 1$ and the average SINR approximates to its maximum value.

3. *Conventional Cellular Networks.* Setting $\theta = 1$ and $K = 1$, the capacity of conventional cellular networks is obtained as $C = \frac{\lambda_1 B}{\lambda_u} \cdot R_{e\_i}$, where $R_{e\_i} =$

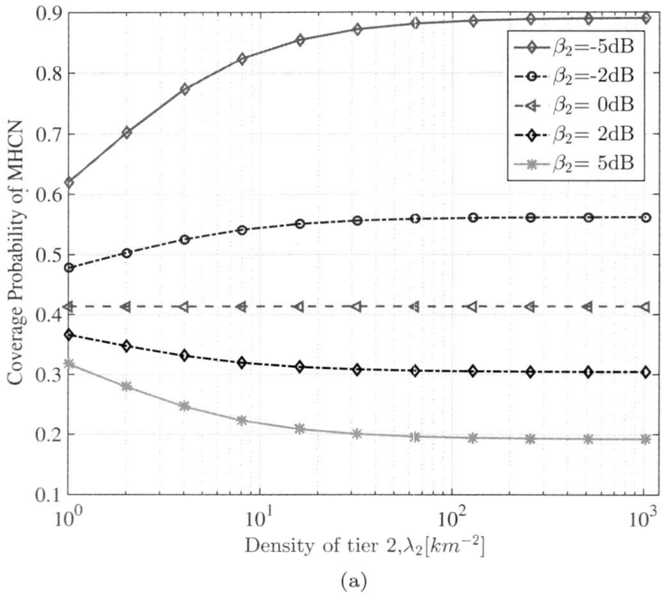

(a)

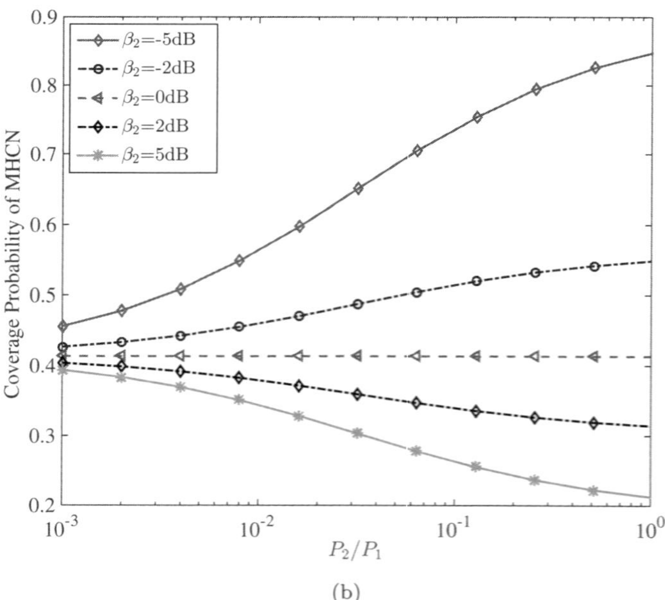

(b)

Figure 2.2: Coverage probability of the two-tier MHCN as a function of λ_2 or P_2/P_1, (a) for BS density of tier two λ_2 with $P_1 = 1.5P_2 = 30P_u$, $\lambda_1 = 1\ km^{-2}$, $\beta_1 = \beta_u = 0$ dB, $\alpha = 3$), and (b) for power ratio P_2/P_1 with $P_1 = 46$ dBm, $P_u = 10$ dBm, $\lambda_u = 10\lambda_2 = 100\lambda_1$.

$\log\left(1 + \beta_{min}\right) + \frac{\sum_{i=1}^{K} \lambda_i P_i^{2/\alpha} A(\alpha,\beta_i,\beta_{min})}{\sum_{i=1}^{K} \lambda_i P_i^{2/\alpha} \beta_i^{-2/\alpha}}$. Compared with conventional cellular network, the capacity gain of MHCNs comes from two aspects: the enhancement of spatial reuse and the reduction of single-hop transmission distance.

2.1.5　Numerical Analysis and Discussions

In this section, we will obtain the numerical results of coverage probability and capacity of MHCNs with respect to main system parameters, such as BS density, transmit power and SINR threshold. Specifically, we use the two-tier MHCN as an example.

Figure 2.2 shows the coverage probability of the two-tier MHCN as a function of the BS density (Figure 2.2(a)) or the transmit power of tier 2 (Figure 2.2(b)). With the above analysis in this chapter, we find that when $K = 2$, the SIFigureNR enhancement condition is simplified to $\beta_2 < \beta_1$. We observe that when $\beta_2 = -5$ dB or $\beta_2 = -2$ dB, the coverage probability of MHCN ascends with the BS density and transmit power of tier 2 since the received signal power is larger than the increased interference power. In the case of $\beta_2 = 0$ dB, the coverage probability remains invariant with the change of BS density and the transmit power of tier 2 because the received SINR distribution is unchanged with the above parameters. When $\beta_2 = 2$ dB or $\beta_2 = 5$ dB, the SINR enhancement condition is dissatisfied, and the increase in BS density and transmit power leads to a decrease in the coverage probability because of severe intercell interference.

Figure 2.3 shows the capacity of MHCN as a function of the BS density (Figure 2.3(a)) or the transmit pwoer of tier 2 (Figure 2.3(b)) where we vary the parameters θ and β_2. In both figures, we observe that the network capacity decreases with the decreasing θ, which means the division of resources between direct links and ad hoc links diminishes the capacity. In addition, Figure 2.3(a) shows that the capacity of MHCN increases with the BS density regardless whether the SE condition is satisfied. This can be explained by the fact that the capacity gain achieved by the increase of BS density λ_2 has an advantage over the reducing spectral efficiency caused by the decreased SIR threshold β_2 , thus dominating the tendency of network capacity. In Figure 2.3(b), the network capacity increases with P_2 when $\beta_2 = 5$ dB but decreases with P_2 when $\beta_2 = -5$ dB. When $\beta_2 = 5$ dB, the SE enhancement condition is satisfied, thus the increase of transmit power will improve the spectral efficiency, thereby enhancing the capacity of MHCNs while the opposite occurs when $\beta_2 = -5$ dB. With the analytical framework derived in this chapter, we can properly set system parameters to satisfy the SE condition to maximize network capacity.

2.1.6　Conclusions

In this chapter, we analyzed the capacity of MHCNs where the low-power BSs are allocated overlaid with the traditional macrocells and the data generated from BSs is transmitted to their destinations via one hop or two hops. The

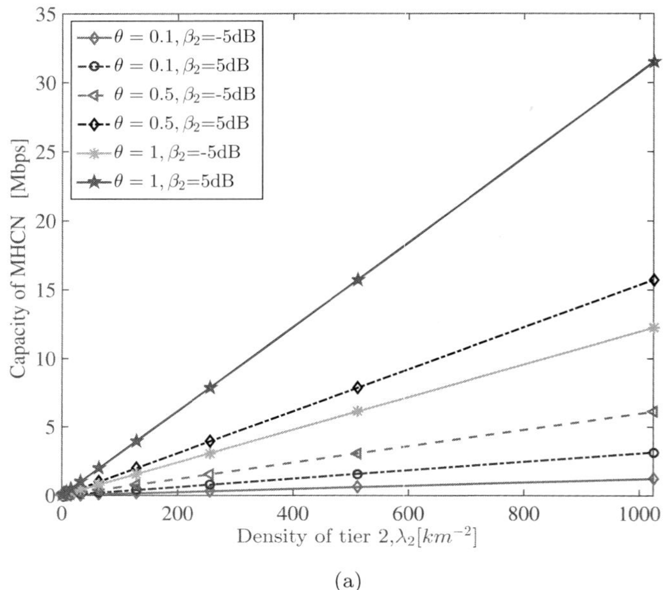

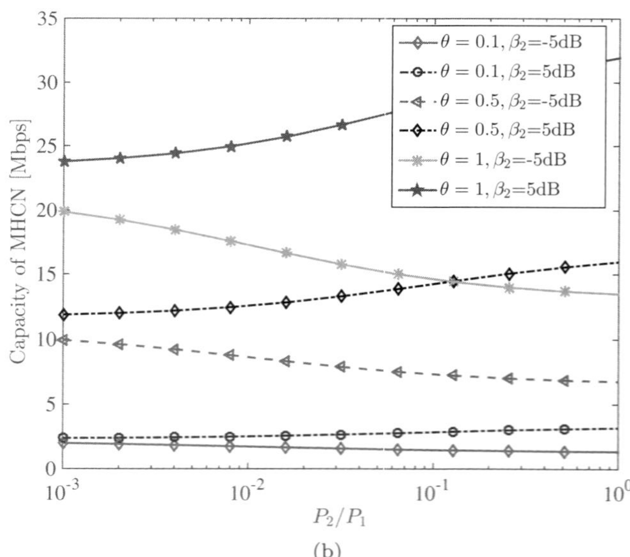

Figure 2.3: Capacity of the two-tier MHCN as a function of λ_2 or P_2/P_1 with $\beta_1 = \beta_u = 0$ dB, B=10 MHz, $\alpha = 3$, (a) for BS density of tier 2 λ_2 with $P_1 = 1.5P_2 = 30P_u$, $\lambda_1 = 1\ km^{-2}$, and (b) for power ratio P_2/P_1 with $P_1 = 46$ dBm, $P_u = 10$ dBm, $\lambda_u = 10\lambda_2 = 100\lambda_1$.

frequency bands used by the transmissions from BSs to users are different than those from users to users. Provided that the different classes of BSs and the mobile users are independent homogeneous PPPs, we derived the expression of the capacity. Compared with traditional cellular networks, the capacity gain of MHCNs comes from the improvement of spatial reuse due to cell splitting and the reduction of path loss because of the ad hoc access mode. Some important conclusions were obtained as follows: (1) When the given SINR thresholds of various tiers are not all the same, the capacity can be enhanced by increasing the BS density and transmit power of a certain tier under the spectral efficiency enhancement condition. Due to the cell splitting, BS density has a better effect on capacity enhancement than transmit power. (2) When all the tiers have the same SINR threshold, the capacity increases linearly with BS density since the spacial reuse is improved. Moreover, the capacity increases with SINR threshold.

2.2 D2D Enhanced Cellular Networks: Spectrum Sharing Schemes

2.2.1 Introduction

D2D communications are seen as a promising technique to meet the emerging proximity-based services, such as local advertising and social networking applications. By establishing a direct link between UEs, D2D communication leads to three potential advantages: increased spectral efficiency, low energy consumption, and reduced end-to-end latency [24, 25]. Meanwhile, the HCN has been shown to be a cost-effective means to enhance network capacity and coverage probability by combining the conventional macrocell network with diverse low-power SAPs [26, 27]. As such, spectrum sharing between heterogeneous networks and D2D communications is attracting more and more attention. With tools from stochastic geometry [28], tractable analytical frameworks were developed for the design of D2D spectrum sharing in frequency-division duplex (FDD) cellular networks [29–32]. Game theoretical models were applied to study resource allocation for D2D communication in [25, 33, 34].

Compared with FDD, which requires two symmetrical frequency bands for the uplink (UL) and downlink (DL), TDD only uses a single frequency band for transmission and receive [35]. Aside from spectrum saving, another advantage of TDD is its ability to accommodate the asymmetric traffic loads in UL and DL by adjusting the fraction of time slots allocated to UL and DL transmissions. Recently, to cope with the instantaneous traffic variations in UL and DL and among different cells, the 3GPP is now introducing the concept of dynamic TDD. In contrast with the semistatic TDD, where all cells employ the same UL/DL configuration, dynamic TDD allows a more flexible use of cell-specific UL/DL configuration [36, 37]. Specifically, a cell determines its own UL/DL configuration based on its instantaneous traffic condition and can further reset the configuration in a much faster timescale. As such, network resources are better utilized and network performance is enhanced in terms of network throughput and end-to-end latency.

In a two-tier HCN operating with universal frequency reuse, the major challenge is the cross-tier and cotier interference. In addition, extra interference is imposed to the cellular transmissions if underlaid D2D transmissions are admitted. In a network operating with dynamic TDD, interference becomes more severe due to new generated cross-subframe interference, namely, mobile user-to-mobile user (UL-to-DL) interference and base station-to-base station (DL-to-UL) interference when the neighboring cells operate in opposite directions. It has been proven that DL-to-UL interference may lead to an unacceptable performance in UL transmissions [38]. Thus, to fully exploit the gain from dynamic TDD and D2D communication, interference management strategies must be considered. To alleviate the cross-tier interference in a two-tier HCN, variable interference management schemes have been proposed, such as power control [39], interference cancellation [40], and spectrum allocation [41]. For spectrum allocation, a distributed disjoint subchannel allocation policy is sensible, especially in dense small cell networks [41]. To address the cotier interference, medium access control (MAC) is an effective and widely used technique in distributed ad hoc/sensor networks [42–44]. Carrier sensing multiple access (CSMA) is a popular MAC protocol where the positions of simultaneously transmitting nodes can be modeled by a *Matern hard-core process (MHP)* [42,43,45]. In an MHP, each node respects a minimum exclusion distance with respect to each other to control the mutual interference. Carrier sensing is also employed in cognitive radio networks to limit the interference inflicted on primary users (PUs). In [44], secondary users (SUs) are modeled as a *Poisson hole process (PHP)*, so that only SUs located outside the exclusion region of PUs can transmit.

Despite the fact that both the merit of dynamic TDD networks [36, 37] and the benefit of D2D communications in FDD networks [29–31] have been widely discussed in the literature, a unifying framework for D2D-enhanced TDD networks is still missing. In this chapter, we present an analytical framework to evaluate network performance in terms of *load-aware* coverage probability and network throughput. The proposed framework allows us to quantify the effect on the coverage probability of the most important TDD system parameters, such as UL/DL configuration, base station density, and bias factor. In addition, we evaluate how the bandwidth partition and the D2D network access scheme affect total network throughput. Through the study of the trade-off between coverage probability and D2D user activity, we provide guidelines for the optimal design of D2D network access.

2.2.2   D2D Enhanced HCN

Let us consider a two-tier HCN consisting first of a tier of macro base stations (MBSs) overlaid with a network of SAPs. With tools from stochastic geometry, we model the spatial distribution of MBSs and SAPs as homogeneous poisson point processes (PPPs). Specifically, MBSs are distributed according to a homogeneous PPP Φ_m with spatial density λ_m, and SAPs are distributed according to a PPP Φ_s with density λ_s. Mobile users are scattered over $\mathbb{R}^2$ according to a PPP Φ_u with spatial density λ_u. To model the D2D communications, we assume that a fraction ζ of the mobile users have their target receiver within a close distance and are considered as potential D2D transmitters. With the thinning theorem, the set of potential D2D transmitters $\tilde{\Phi}_d = \{T_i\}$ forms a PPP with density $\zeta\lambda_u$. Each potential D2D transmitter is assumed to have an assigned receiver (not belonging to Φ_u) at a fixed distance r_d in a uniformly random direction. We note that the potential D2D receivers are scattered according to a PPP with density $\zeta\lambda_u$, where the potential D2D receivers and Φ_u are dependent point processes. To alleviate network interference, orthogonal spectrum allocation is considered between the two tiers. The total bandwidth W is divided into two nonoverlapping parts ηW and $(1 - \eta)W$ that are allocated to the macrotier and small cell tier. With this assumption, cross-tier interference is eliminated. We let potential D2D users share the spectrum with the small cell tier, which leads to interference between small cell users and D2D users.

We consider a dynamic TDD scheme for both the macrotier and small cell tier, where at each time slot a cell configures flexibly in DL or UL mode. The transmission mode selection for macrocells and small cells is modeled by independent Bernoulli random variables (r.v.s), so that macrocells and small cells are configured in DL mode with probability $q_{D,m}$ and $q_{D,s}$, respectively, while the corresponding UL mode probabilities are given by $1 - q_{D,m}$ and $1 - q_{D,s}$. The multiplexing probabilities $q_{D,m}$, and $q_{D,s}$ define the UL/DL configuration for the macrotier and small cell tier. The concurrent DL and UL transmissions in neighboring cells may lead to new types of intercell interference, i.e. DL-to-UL and UL-to-DL (user-to-user) interference. Let P_m, P_s, Q_m and Q_s denote the transmit power of MBSs, SAPs, mobile users associated with the macrotier, and mobile users associated with the small cell tier. We consider a baseline model that does not account for UL power control. However, it is possible to extend the network performance analysis to the case with power control policy by applying the results developed in [29, 32, 46]. We use Q_d to represent the transmit power of potential D2D users. To avoid confusion, in the following parts the term *mobile users* only denotes mobile users that are expected to communicate via infrastructures, while the term *potential D2D users* refers to mobile users that attempt to communicate with each other by employing D2D technology.

In the network, we employ a load-aware resource allocation model where each base station always has data to transmit if it has a mobile user within its coverage. The orthogonal multiple access scheme is considered, so that at any given time slot and subchannel only a single mobile user can be active in a cell. If several mobile users connect to the same base station, the base station will randomly choose one mobile user to serve. Note that due to spectrum sharing between D2D users and small cell users, the cochannel interference is introduced. To control the interference, we propose a CSMA scheme to limit the activity of D2D transmitters. The channel between any pair of users is assumed to be independent and consists of large-scaled path loss and small-scaled Rayleigh fading. In the analysis, we ignore the thermal noise and consider the interference-limited regime. This assumption is reasonable in the interference-dominated heterogeneous networks [47].

2.2.3   A Decoupled Cell Association Scheme

For the traditional reference signal received power (RSRP) cell selection policy, the mobile user connects to the base station with the strongest DL received power. However, the great disparity between P_m and P_s leads to larger coverage area of the macrocell and smaller coverage area of the small cell. This leads to the fact that a mobile user is geographically closer to a SAP but has to connect to the MBS that has a larger geographical distance to the mobile user. In this way, mobile users prone to the MBSs and leave the abundant resources of SAPs underutilized, causing load imbalances. In addition, the RSRP cell selection policy is nonideal for UL association. This because in UL, the radio condition of a mobile user is mainly determined by the pathloss to the serving base station, which has direct relation with the geographical distance. To cope with the large pathloss in UL, the macro mobile user has to transmit with high power. However, the high transmit power imposes strong cross-tier interference to the UL SAPs nearby, which is denoted by the dotted line in Figure 2.4.

The idea of cell range expansion (CRE) is to extend the coverage area of a small cell by adding a positive bias factor B_s to P_s, so that mobile users are artificially biased to the low-power SAPs. As a result, parts of mobile users are offloaded from MBSs to nearby SAPs, making the traffic load between the two tiers more balanced. As shown in Figure 2.4, by using CRE, the mobile user is shifted from the MBS to the nearby SAP and then the cross-tier interference from the macro mobile user to the SAP is mitigated. Also, by associating with the SAP, the mobile user can transmit with low power, leading to less cross-tier interference to the MBSs. However, since the mobile user locates in the range expanded area in DL case, it suffers from strong interference from the MBS and hence results in bad signal conditions at the mobile user location.

Besides CRE, another technique to mitigate cross-tier interference in HCN is almost blank subframe (ABS) which serves as an important part of the enhanced intercell interference coordination (eICIC) framework in 3GPP. With this technique, some subframes of the macrocell is blanked and left for scheduling vulnerable small cell users. It is worth noting that there are still some resource elements that keep transmitting some specific signals. For example, cell-specific reference symbols (CRS) are used for doing channel estimation and radio resource management.

In this chapter, the motivation of proposing a decoupled cell association scheme is the drawback of the CRE technique. Note that the non-ideal UL association is caused by the fact that RSRP cell selection policy is employed in both DL and UL transmissions. An alternative way to alleviate the problem is to explicitly define different association policies in UL and DL. In this way, the cell associations in UL and DL are decoupled, leading to great flexibility to deal with the load imbalance problem in each direction. Therefore, in this chapter, we model different associations for UL and DL and define the biased association policy in each direction to make the framework more general. At each time slot, a mobile user acts as a transmitter or receiver with probability μ and $1 - \mu$, respectively. Assuming open access, the association of a mobile user to a given tier is based on the maximum biased received signal power averaged over fading. The bias factor in this association policy is used to balance the traffic load among different tiers. In this chapter, the decoupled DL and UL association model is given as follows.

A typical receiver is associated with the nearest base station in DL mode of tier i if

$$i = \arg \max_{k \in \{\mathtt{m},\,\mathtt{s}\}} P_k B_{\mathrm{D},k} D_{\mathrm{D},k}^{-\alpha},$$

where $B_{\mathrm{D},k}$ is the DL bias factor of tier k, and $D_{\mathrm{D},k}$ denotes the distance from the typical receiver to the nearest base station of Φ_k operating in DL mode with thinned density $q_{\mathrm{D},k}\lambda_k$.

A typical transmitter is associated with the nearest base station in UL mode of tier i if

$$i = \arg \max_{k \in \{\mathtt{m},\,\mathtt{s}\}} Q_k B_{\mathrm{U},k} D_{\mathrm{U},k}^{-\alpha},$$

where $B_{\mathrm{U},k}$ is the UL bias factor of tier k, $D_{\mathrm{U},k}$ denotes the distance from the typical transmitter to the nearest base station of Φ_k operating in UL mode with thinned density $(1 - q_{\mathrm{D},k}) \lambda_k$.

For notational brevity, we define several normalized parameters of tier k conditioned on the serving tier i: $\hat{\lambda}_k^{(i)} \triangleq \frac{\lambda_k}{\lambda_i}$, $\hat{q}_{\mathrm{D},k}^{(i)} \triangleq \frac{q_{\mathrm{D},k}}{q_{\mathrm{D},i}}$, $\hat{q}_{\mathrm{U},k}^{(i)} \triangleq \frac{1-q_{\mathrm{D},k}}{1-q_{\mathrm{D},i}}$, $\hat{P}_k^{(i)} \triangleq \frac{P_k}{P_i}$, $\hat{Q}_k^{(i)} \triangleq \frac{Q_k}{Q_i}$, $\hat{B}_{\mathrm{D},k}^{(i)} \triangleq \frac{B_{\mathrm{D},k}}{B_{\mathrm{D},i}}$, $\hat{B}_{\mathrm{U},k}^{(i)} \triangleq \frac{B_{\mathrm{U},k}}{B_{\mathrm{U},i}}$. Using the proposed association rules, the set of base stations form different multiplicatively weighted Voronoi tessellations of the two-dimensional plane in DL and UL. As a consequence, a mobile user may associate with different base stations for DL and UL traffic. With the proposed decoupled association policy, we derive the probability that a typical receiving and transmitting mobile user is associated with tier i for DL and UL as

$$\mathcal{A}_{\mathrm{D},i} \;=\; \frac{q_{\mathrm{D},i}\lambda_i}{\sum_{k \in \{\mathtt{m},\mathtt{s}\}} G_{\mathrm{D},k}^{(i)}}, \; \mathcal{A}_{\mathrm{U},i} = \frac{(1 - q_{\mathrm{D},i}) \lambda_i}{\sum_{k \in \{\mathtt{m},\mathtt{s}\}} G_{\mathrm{U},k}^{(i)}}, \tag{2.30}$$

where $G_{\mathrm{D},k}^{(i)} = q_{\mathrm{D},k}\lambda_k \left(\hat{P}_k^{(i)} \hat{B}_{\mathrm{D},k}^{(i)} \right)^{\frac{2}{\alpha}}$, $G_{\mathrm{U},k}^{(i)} = (1 - q_{\mathrm{D},k}) \lambda_k \left(\hat{Q}_k^{(i)} \hat{B}_{\mathrm{U},k}^{(i)} \right)^{\frac{2}{\alpha}}$. For the special case of $\{q_{\mathrm{D},\mathtt{m}},$
$q_{\mathrm{D},\mathtt{s}}\} = \{0,0\}$, we define $\{\mathcal{A}_{\mathrm{D},\mathtt{m}}, \mathcal{A}_{\mathrm{D},\mathtt{s}}\} = \{0,0\}$, and the network changes into a two-tier UL network. For $\{q_{\mathrm{D},\mathtt{m}}, q_{\mathrm{D},\mathtt{s}}\} = \{1,1\}$, we define $\{\mathcal{A}_{\mathrm{U},\mathtt{m}}, \mathcal{A}_{\mathrm{U},\mathtt{s}}\} = \{0,0\}$, and the network transforms to a two-tier DL network. The association probabilities defined in (2.30) indicate how the per-tier association probability in a two-tier dynamic TDD network depends on the *relative* transmit power, bias factor, and base station density of the corresponding transmission mode. Note that the base station density affects the per-tier association probability more than transmit power or bias factor. The distance between a typical mobile user

and its serving base station of tier i in DL or UL mode is given by [48]

$$f_{Y_{\mathrm{D},i}}(y) = 2\pi \frac{q_{\mathrm{D},i}\lambda_i}{\mathcal{A}_{\mathrm{D},i}} y \exp\left\{-\pi \frac{q_{\mathrm{D},i}\lambda_i}{\mathcal{A}_{\mathrm{D},i}} y^2\right\}, \tag{2.31}$$

$$f_{Y_{\mathrm{U},i}}(y) = 2\pi \frac{(1 - q_{\mathrm{D},i})\lambda_i}{\mathcal{A}_{\mathrm{U},i}} y \exp\left\{-\pi \frac{(1 - q_{\mathrm{D},i})\lambda_i}{\mathcal{A}_{\mathrm{U},i}} y^2\right\}, \tag{2.32}$$

By considering the traffic load, we derive a more accurate *load-aware* coverage probability. For each tier i, we compute the void probability of a random base station in DL and UL (i.e., $\mathbb{P}_e^{\mathrm{D},i}$ and $\mathbb{P}_e^{\mathrm{U},i}$), and compare it with a threshold value to determine the network traffic load. When $\mathbb{P}_e^{\mathrm{D},i} < 10^{-4}$ and $\mathbb{P}_e^{\mathrm{U},i} < 10^{-4}$, we say tier i is fully loaded, otherwise, it is partially loaded. Denote $\Phi_{\mathrm{m}}^{\mathrm{D}} \sim \mathrm{PPP}(\lambda_{\mathrm{m}}^{\mathrm{D}})$, $\Phi_{\mathrm{s}}^{\mathrm{D}} \sim \mathrm{PPP}(\lambda_{\mathrm{s}}^{\mathrm{D}})$, $\Phi_{\mathrm{m}}^{\mathrm{U}} \sim \mathrm{PPP}(\lambda_{\mathrm{m}}^{\mathrm{U}})$, and $\Phi_{\mathrm{s}}^{\mathrm{U}} \sim \mathrm{PPP}(\lambda_{\mathrm{s}}^{\mathrm{U}})$ as the point processes of *active* DL MBSs, DL SAPs, UL MBSs, and UL SAPs, respectively, with corresponding denstities $\lambda_{\mathrm{m}}^{\mathrm{D}}$, $\lambda_{\mathrm{s}}^{\mathrm{D}}$, $\lambda_{\mathrm{m}}^{\mathrm{U}}$, and $\lambda_{\mathrm{s}}^{\mathrm{U}}$. The DL and UL void probability of a base station in tier i is respectively given by [49]

$$\mathbb{P}_e^{\mathrm{D},i} = \left(1 + \frac{(1-\mu)(1-\zeta)\lambda_{\mathrm{u}}\mathcal{A}_{\mathrm{D},i}}{3.5 q_{\mathrm{D},i}\lambda_i}\right)^{-3.5}, \mathbb{P}_e^{\mathrm{U},i} = \left(1 + \frac{\mu(1-\zeta)\lambda_{\mathrm{u}}\mathcal{A}_{\mathrm{U},i}}{3.5(1 - q_{\mathrm{D},i})\lambda_i}\right)^{-3.5}. \tag{2.33}$$

Furthermore, the density of active base stations in DL and UL mode of tier i is given by $\lambda_i^{\mathrm{D}} = \lambda_i q_{\mathrm{D},i}\left(1 - \mathbb{P}_e^{\mathrm{D},i}\right)$ and $\lambda_i^{\mathrm{U}} = \lambda_i\left(1 - q_{\mathrm{D},i}\right)\left(1 - \mathbb{P}_e^{\mathrm{U},i}\right)$.

2.2.4   CSMA Model for D2D Transmissions

In the beginning of a time slot, each potential D2D transmitter senses the active small cell transmissions that originate from SAPs in DL mode and transmitting mobile users associated with the small cell tier. As such, D2D transmissions respect an exclusion region around each small cell transmitter. The remaining potential D2D transmitters form a PHP, which can be approximated by a PPP [44]. Carrier sensing is also performed with respect to the remaining potential D2D transmitters, where the signal power from a nearby D2D transmitter is not allowed to surpass the contention threshold ρ_{d}. To resolve the collision among the D2D contenders, we use a back-off scheme. Specifically, each remaining D2D transmitter independently samples a random timer $t_i \sim \mathcal{U}[0,1]$ and channel access is granted to the contender with the smallest timer within a contention region [45].

To achieve the channel state information from a potential D2D transmitter to all the active small cell transmitters, we need to add a control channel to support the channel estimation. Taking the TD-LTE standard as an example, we can allocate each small cell transmitter some dedicated resource elements (REs), each of which corresponds to one subcarrier (tone) in the frequency domain and one OFDM symbol in the time domain. Each small cell transmitter can send signals at the dedicated REs with known transmit power, which can be detected by the D2D transmitters. To minimize the overhead, we can employ *analog signaling*, and allocate each small cell transmitter a dedicated RE. Each active small cell transmitter lights up (i.e., transmits a proper known power P_o on) the symbol and tone allocated to it. Each potential D2D transmitter senses all the REs lighted by each small cell transmitter and estimates the channel power gain from itself to all the active small cell transmissions. Assuming channel reciprocity, the potential D2D transmitter predicts the *would-be* interference it may impose on the small cell transmitters and refrains from transmitting if the interference exceeds the protection threshold ρ_s. Since the potential D2D transmitter does not need to distinguish the small cell transmitters, it only costs a little overhead to complete the sensing. For instance, assume that the bandwidth occupied by the small cell tier is 5 MHz, which includes 240 REs within one OFDM symbol. Therefore, it can be used to sense more than two hundred active small cell transmitters within one OFDM symbol.

We define U_i as the retention indicator of the ith potential D2D transmitter T_i, which is given by

$$
\begin{aligned}
U_i \;=\; & \prod_{Y_j \in \Phi_\mathrm{s}^\mathrm{D}} \mathbf{1}_{\left(\frac{Q_\mathrm{d} h_{ji}}{\|T_i - Y_j\|^\alpha} < \rho_\mathrm{s} \right)} \prod_{Z_l \in \Phi_{\mathrm{u,s}}^\mathrm{T}} \mathbf{1}_{\left(\frac{Q_\mathrm{d} h_{li}}{\|T_i - Z_l\|^\alpha} < \rho_\mathrm{s} \right)} \\
& \times \prod_{T_k \in \Phi_\mathrm{d} \backslash T_i} \left(\mathbf{1}_{(t_i \le t_k)} + \mathbf{1}_{(t_i > t_k)} \mathbf{1}_{\left(\frac{Q_\mathrm{d} h_{ki}}{\|T_i - T_k\|^\alpha} < \rho_\mathrm{d} \right)} \right)
\end{aligned}
\tag{2.34}
$$

where h_{ji} denotes the channel fading from T_i to Y_j, $\{Y_j\} = \Phi_\mathrm{s}^\mathrm{D}$ denotes the set of active SAPs in DL, and $\{Z_l\} = \Phi_{\mathrm{u,s}}^\mathrm{T}$ represents the set of transmitting mobile users associated with the small cell tier. The first two products in (2.34) reflect that the interference inflicted by a potential D2D transmitter on an active small cell transmitter should be smaller than ρ_s. The first term inside the last product corresponds to the event where the timer t_i is smaller than t_k, while the second term corresponds to the event where the timer t_i is larger than t_k, yet the interference from T_i to T_k is smaller than ρ_d.

We define $\beta \triangleq \Pr[U_i = 1]$ as the retaining probability of T_i, which depends on the timer t_i, the channel fading and the distance between T_i and small cell transmitters. The set of winning contenders forms a point process similar to the MHP, where any two points respect a minimum exclusion distance determined by ρ_d and the instantaneous channel gain. A node is forbidden to transmit only if there is another node within a certain exclusion distance with a smaller mark [42], since the aggregate interference experienced by a user of an MHP can be approximated by the interference resulting from a PPP that has the same density as the MHP and exists outside the exclusion region [42, 43]. In this work, we assume that each potential D2D transmitter is retained independently with the probability β. As a result, the retained D2D transmitters Φ_d form a PPP with density $\beta \zeta \lambda_\mathrm{u}$, where the combined effect of PHP and MHP is captured by β. Note that the retained D2D transmitters are actually the *active* D2D transmitters in the current time slot. The potential D2D transmitters that fail to access a channel will keep silent in the current time slot and continue to execute the CSMA scheme in the next time slot. With the classical stochastic geometry

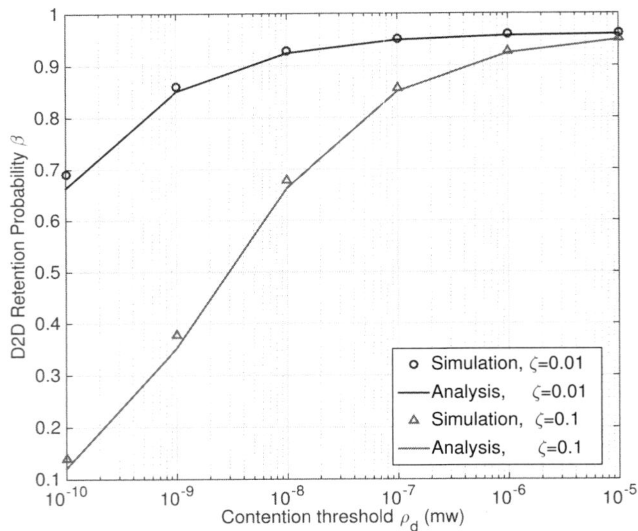

Figure 2.4: D2D retention probability β from simulation (markers) and theoretical analysis (lines) as a function of ρ_d, for $\lambda_s = 5\lambda_m$, $\lambda_u = 100\lambda_m$, $\{q_{D,m}, q_{D,s}\} = \{0.5, 0.5\}$, $\rho_s = -60$ dBm, and $\{B_{D,m}, B_{D,s}, B_{U,m}, B_{U,s}\} = \{1, 1, 1, 1\}$).

analysis, we derive the retaining probability of a potential D2D transmitter as [48]

$$\beta = \exp\left(-\left(\lambda_s^D + \lambda_{u,s}^T\right)\mathcal{K}_{o,s}\right)\frac{1 - \exp\left(-\zeta\lambda_u\mathcal{K}_{o,d}\right)}{\zeta\lambda_u\mathcal{K}_{o,d}}, \qquad (2.35)$$

and the corresponding density of active D2D transmitters is derived as $\lambda_d = \exp\left(-\left(\lambda_s^D + \lambda_{u,s}^T\right)\mathcal{K}_{o,s}\right)$

$\times \frac{1-\exp(-\zeta\lambda_u\mathcal{K}_{o,d})}{\mathcal{K}_{o,d}}$, where $\mathcal{K}_{o,s} = \dfrac{2\pi\Gamma\left(\frac{2}{\alpha}\right)}{\alpha\left(\frac{\rho_s}{Q_d}\right)^{\frac{2}{\alpha}}}$ and $\mathcal{K}_{o,d} = \dfrac{2\pi\Gamma\left(\frac{2}{\alpha}\right)}{\alpha\left(\frac{\rho_d}{Q_d}\right)^{\frac{2}{\alpha}}}$. Figure 2.4 shows the D2D retention probability β as a function of D2D contention threshold ρ_d. We observe a good match between simulations and analytical results. Low values of ρ_d correspond to large exclusion distances ι_d, leading to a low retaining probability β.

2.2.5   Coverage Probability

By definition, the load-aware coverage probability of tier i, $i \in \{m, s\}$ in DL and UL is given by

$$\mathbb{P}_i^D = \Pr[\text{SIR}_i^D > \gamma_i^D], \quad \mathbb{P}_i^U = \Pr[\text{SIR}_i^U > \gamma_i^U], \qquad (2.36)$$

where γ_i^D and γ_i^U denote the SIR thresholds of DL and UL transmissions in tier i. Similarly, the coverage probability of a typical D2D receiver is $\mathbb{P}_d = \Pr[\text{SIR}_d > \gamma_d]$ with γ_d the SIR threshold of D2D user.

The DL and UL SIR of a typical receiver associated with the macro tier is given by

$$\text{SIR}^{\text{D}}_{\text{m}} = \frac{P_{\text{m}} h_{or} r^{-\alpha}}{I^{(\text{m})}_{\text{D}\to\text{D}} + I^{(\text{m})}_{\text{U}\to\text{D}}}, \quad \text{SIR}^{\text{U}}_{\text{m}} = \frac{Q_{\text{m}} h_{or} r^{-\alpha}}{I^{(\text{m})}_{\text{D}\to\text{U}} + I^{(\text{m})}_{\text{U}\to\text{U}}}, \tag{2.37}$$

where h_{or} and r are the fading power and the typical link length, and

$$I^{(\text{m})}_{\text{D}\to\text{D}} = \sum_{y\in\Phi^{\text{D}}_{\text{m}}\setminus\{y_0\}} P_{\text{m}} h_{oy} y^{-\alpha}, \quad I^{(\text{m})}_{\text{U}\to\text{D}} = \sum_{x\in\Phi^{\text{T}}_{\text{u,m}}} Q_{\text{m}} h_{ox} x^{-\alpha},$$

$$I^{(\text{m})}_{\text{D}\to\text{U}} = \sum_{y\in\Phi^{\text{D}}_{\text{m}}} P_{\text{m}} h_{oy} y^{-\alpha}, \quad I^{(\text{m})}_{\text{U}\to\text{U}} = \sum_{x\in\Phi^{\text{T}}_{\text{u,m}}\setminus\{x_0\}} Q_{\text{m}} h_{ox} x^{-\alpha},$$

where y_0 and x_0 represent the position of typical transmitter in DL and UL mode and $\Phi^{\text{T}}_{\text{u,m}}$ represents the set of transmitting mobile users associated with macrotier. Due to the orthogonal multiple access technology, there is a one-to-one mapping from the transmitting mobile users associated with the macrotier to the active UL MBSs. Since the coupling between the location of MBSs and transmitting mobile users has little effect on the coverage probability [46, 50], we neglect the coupling and model $\Phi^{\text{T}}_{\text{u,m}}$ as a PPP with density $\lambda^{\text{T}}_{\text{u,m}} = \lambda^{\text{U}}_{\text{m}}$.

The DL and UL SIR of a typical receiver associated with small cell tier is denoted by

$$\text{SIR}^{\text{D}}_{\text{s}} = \frac{P_{\text{s}} h_{or} r^{-\alpha}}{I^{(\text{s})}_{\text{D}\to\text{D}} + I^{(\text{s})}_{\text{U}\to\text{D}} + I_{\text{d}\to\text{D}}}, \quad \text{SIR}^{\text{U}}_{\text{s}} = \frac{Q_{\text{s}} h_{or} r^{-\alpha}}{I^{(\text{s})}_{\text{D}\to\text{U}} + I^{(\text{s})}_{\text{U}\to\text{U}} + I_{\text{d}\to\text{U}}},$$

$$\tag{2.38}$$

where

$$I^{(\text{s})}_{\text{D}\to\text{D}} = \sum_{y\in\Phi^{\text{D}}_{\text{s}}\setminus\{y_0\}} P_{\text{s}} h_{oy} y^{-\alpha}, \quad I^{(\text{s})}_{\text{U}\to\text{D}} = \sum_{x\in\Phi^{\text{T}}_{\text{u,s}}} Q_{\text{s}} h_{ox} x^{-\alpha}, \quad I_{\text{d}\to\text{D}} = \sum_{z\in\Phi_{\text{d}}\setminus b(y_0,\iota_{\text{s}})} Q_{\text{d}} h_{oz} z^{-\alpha},$$

$$I^{(\text{s})}_{\text{D}\to\text{U}} = \sum_{y\in\Phi^{\text{D}}_{\text{s}}} P_{\text{s}} h_{oy} y^{-\alpha}, \quad I^{(\text{s})}_{\text{U}\to\text{U}} = \sum_{x\in\Phi^{\text{T}}_{\text{u,s}}\setminus\{x_0\}} Q_{\text{s}} h_{ox} x^{-\alpha}, \quad I_{\text{d}\to\text{U}} = \sum_{z\in\Phi_{\text{d}}\setminus b(x_0,\iota_{\text{s}})} Q_{\text{d}} h_{oz} z^{-\alpha},$$

where $\Phi^{\text{T}}_{\text{u,s}} \sim \text{PPP}(\lambda^{\text{T}}_{\text{u,s}})$ represents the set of transmitting mobile users associated with small cell tier with density $\lambda^{\text{T}}_{\text{u,s}} = \lambda^{\text{U}}_{\text{s}}$. Note that in the analysis, we approximate the exclusion region of an active D2D transmitter with a ball $\mathcal{B} \triangleq b(y_0, \iota_{\text{s}})$. In fact, due to channel fading, the exclusion region for D2D transmissions around each small cell transmitter is not a ball but an irregular shape. As is illustrated in Figure 2.5, we put a small cell transmitter, such as a DL SAP, at the center, and the irregular dotted line is the boundary of the exclusion region. When the small cell transmitter transmits with power Q_d, the received signal power on each point of the dotted line is equal to ϱs. The potential D2D transmitters within the exclusion region, denoted by the black nodes, cannot be active, while the nodes outside the dotted line are the remaining active D2D transmitters. Note that the irregular shape of the exclusion region makes the

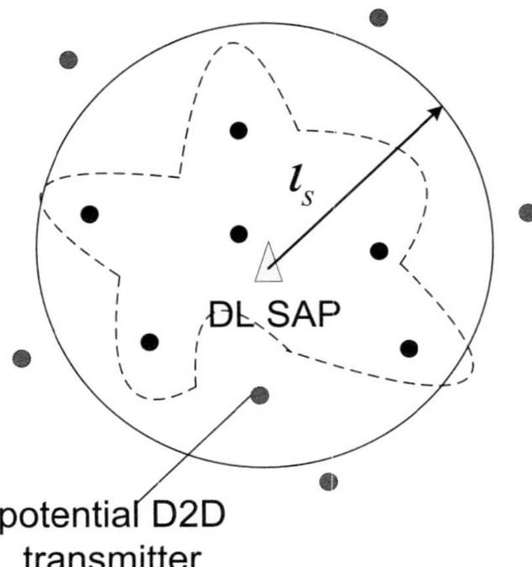

Figure 2.5: Illustration of the irregular shape of the exclusion region, which is approximated by a ball.

analysis difficult. To simplify the analysis, we use a circle, represented by the solid line, to approximate the irregular shape. The equivalent exclusion distance is determined by imposing a small misdetection rate threshold ϵ. The constraint is met when $\Pr[Q_\mathrm{d} h_{x_0 z}/\iota_\mathrm{s}^\alpha > \rho_\mathrm{s}] = \epsilon$, and solving for ι_s yields $\iota_\mathrm{s} = \left(\frac{-\ln\epsilon}{\rho_\mathrm{s}/Q_\mathrm{d}}\right)^{\frac{1}{\alpha}}$, where $h_{x_0 z}$ denotes the fading power from a D2D transmitter z to the typical small cell transmitter x_0.

The SIR of a typical D2D receiver is given by

$$\mathrm{SIR_d} = \frac{Q_\mathrm{d} h_{or} r_\mathrm{d}^{-\alpha}}{I_{\mathrm{D}\to\mathrm{d}}^{(\mathrm{s})} + I_{\mathrm{U}\to\mathrm{d}}^{(\mathrm{s})} + I_{\mathrm{d}\to\mathrm{d}}}, \tag{2.39}$$

where

$$I_{\mathrm{D}\to\mathrm{d}}^{(\mathrm{s})} = \sum_{y\in\Phi_\mathrm{s}^\mathrm{D}\setminus b(z_0,\iota_\mathrm{s})} P_\mathrm{s} h_{oy} y^{-\alpha}, \ I_{\mathrm{U}\to\mathrm{d}}^{(\mathrm{s})}$$

$$= \sum_{x\in\Phi_{\mathrm{u},\mathrm{s}}^\mathrm{T}\setminus b(z_0,\iota_\mathrm{s})} Q_\mathrm{s} h_{ox} x^{-\alpha}, \ I_{\mathrm{d}\to\mathrm{d}} = \sum_{z\in\Phi_\mathrm{d}\setminus b(z_0,\iota_\mathrm{d})} Q_\mathrm{d} h_{oz} z^{-\alpha}.$$

There are two exclusion regions $b(z_0,\iota_\mathrm{s})$ and $b(z_0,\iota_\mathrm{d})$ around each retained D2D transmitter z_0, where the former is due to the sensing for small cell transmissions, and the latter results from the sensing among D2D transmitters. Similar to the approximation for the exclusion region around each small cell transmitter, the exclusion regions around each retained D2D transmitter are also approximated by two concentric balls with radius ι_s and ι_d, respectively. The radius ι_d is constrained by ϵ as $\Pr[Q_\mathrm{d} h_{z_0 z}/\iota_\mathrm{d}^\alpha > \rho_\mathrm{d}] = \epsilon$, and solving for ι_d yields $\iota_\mathrm{d} = \left(\frac{-\ln\epsilon}{\rho_\mathrm{d}/Q_\mathrm{d}}\right)^{\frac{1}{\alpha}}$.

In a two-tier dynamic TDD HCN, the overall load-aware coverage probability of a mobile user associated with the infrastructure in DL and UL mode is given by [48]

$$\bar{\mathbb{P}}_{\mathrm{D}} = \mathbb{P}_{\mathrm{m}}^{\mathrm{D}}\mathcal{A}_{\mathrm{D,m}} + \mathbb{P}_{\mathrm{s}}^{\mathrm{D}}\mathcal{A}_{\mathrm{D,s}}, \ \bar{\mathbb{P}}_{\mathrm{U}} = \mathbb{P}_{\mathrm{m}}^{\mathrm{U}}\mathcal{A}_{\mathrm{U,m}} + \mathbb{P}_{\mathrm{s}}^{\mathrm{U}}\mathcal{A}_{\mathrm{U,s}}, \qquad (2.40)$$

and the coverage probability of the typical D2D receiver is derived as

$$\mathbb{P}_{\mathrm{d}} = \exp\left(-\mathcal{I}_1\left(\frac{\gamma_{\mathrm{d}} r_{\mathrm{d}}^{\alpha}}{Q_{\mathrm{d}}}; P_{\mathrm{s}}\right) - \mathcal{I}_2\left(\frac{\gamma_{\mathrm{d}} r_{\mathrm{d}}^{\alpha}}{Q_{\mathrm{d}}}; Q_{\mathrm{s}}\right) - \mathcal{I}_3\left(\frac{\gamma_{\mathrm{d}} r_{\mathrm{d}}^{\alpha}}{Q_{\mathrm{d}}}; Q_{\mathrm{d}}\right)\right), \qquad (2.41)$$

where

$$\mathbb{P}_{\mathrm{m}}^{\mathrm{D}} = \frac{q_{\mathrm{D,m}}\lambda_{\mathrm{m}}}{\lambda_{\mathrm{m}}^{\mathrm{D}}\mathcal{A}_{\mathrm{D,m}}\delta\left(\gamma_{\mathrm{m}}^{\mathrm{D}}, \alpha\right) + \lambda_{\mathrm{u,m}}^{\mathrm{T}}\mathcal{A}_{\mathrm{D,m}}C\left(\alpha\right)\left(\frac{Q_{\mathrm{m}}}{P_{\mathrm{m}}}\gamma_{\mathrm{m}}^{\mathrm{D}}\right)^{\frac{2}{\alpha}} + q_{\mathrm{D,m}}\lambda_{\mathrm{m}}}, \qquad (2.42)$$

$$\mathbb{P}_{\mathrm{m}}^{\mathrm{U}} = \frac{(1 - q_{\mathrm{D,m}})\lambda_{\mathrm{m}}}{C\left(\alpha\right)\left(\gamma_{\mathrm{m}}^{\mathrm{U}}\right)^{\frac{2}{\alpha}}\mathcal{A}_{\mathrm{U,m}}\left(\lambda_{\mathrm{m}}^{\mathrm{D}}\left(\frac{P_{\mathrm{m}}}{Q_{\mathrm{m}}}\right)^{\frac{2}{\alpha}} + \lambda_{\mathrm{u,m}}^{\mathrm{T}}\right) + (1 - q_{\mathrm{D,m}})\lambda_{\mathrm{m}}}, \qquad (2.43)$$

$$\mathbb{P}_{\mathrm{s}}^{\mathrm{D}} = \frac{\pi q_{\mathrm{D,s}}\lambda_{\mathrm{s}}}{\mathcal{A}_{\mathrm{D,s}}}\left[\int_0^{\iota_{\mathrm{s}}^2} e^{-\pi v \mathcal{F}}\mathcal{L}_{I_{\mathrm{d}\to\mathrm{D}}}\left(\frac{\gamma_{\mathrm{s}}^{\mathrm{D}} v^{\frac{\alpha}{2}}}{P_{\mathrm{s}}} \mid v \leq \iota_{\mathrm{s}}^2\right)dv \right.$$
$$\left. + \int_{\iota_{\mathrm{s}}^2}^{\infty} e^{-\pi v \mathcal{F}}\mathcal{L}_{I_{\mathrm{d}\to\mathrm{D}}}\left(\frac{\gamma_{\mathrm{s}}^{\mathrm{D}} v^{\frac{\alpha}{2}}}{P_{\mathrm{s}}} \mid v > \iota_{\mathrm{s}}^2\right)dv\right], \qquad (2.44)$$

$$\mathbb{P}_{\mathrm{s}}^{\mathrm{U}} = \frac{\pi(1 - q_{\mathrm{D,s}})\lambda_{\mathrm{s}}}{\mathcal{A}_{\mathrm{U,s}}}\left[\int_0^{\iota_{\mathrm{s}}^2} e^{-\pi v \mathcal{G}}\mathcal{L}_{I_{\mathrm{d}\to\mathrm{U}}}\left(\frac{\gamma_{\mathrm{s}}^{\mathrm{U}} v^{\frac{\alpha}{2}}}{Q_{\mathrm{s}}} \mid v \leq \iota_{\mathrm{s}}^2\right)dv \right.$$
$$\left. + \int_{\iota_{\mathrm{s}}^2}^{\infty} e^{-\pi v \mathcal{G}}\mathcal{L}_{I_{\mathrm{d}\to\mathrm{U}}}\left(\frac{\gamma_{\mathrm{s}}^{\mathrm{U}} v^{\frac{\alpha}{2}}}{Q_{\mathrm{s}}} \mid v > \iota_{\mathrm{s}}^2\right)dv\right], \qquad (2.45)$$

$$\mathcal{I}_1\left(s; P_{\mathrm{s}}\right) = \pi \lambda_{\mathrm{s}}^{\mathrm{D}}\left(\iota_{\mathrm{s}} + r_{\mathrm{d}}\right)^2\delta\left(\frac{sP_{\mathrm{s}}}{\left(\iota_{\mathrm{s}} + r_{\mathrm{d}}\right)^{\alpha}}, \alpha\right) + \lambda_{\mathrm{s}}^{\mathrm{D}}\mathcal{Z}_{0, \iota_{OE}}^{\pi, \iota_{\mathrm{s}} + r_{\mathrm{d}}}\left(s; P_{\mathrm{s}}\right), \ (2.46)$$

$$\mathcal{I}_2\left(s; Q_{\mathrm{s}}\right) = \pi \lambda_{\mathrm{u,s}}^{\mathrm{T}}\left(\iota_{\mathrm{s}} + r_{\mathrm{d}}\right)^2\delta\left(\frac{sQ_{\mathrm{s}}}{\left(\iota_{\mathrm{s}} + r_{\mathrm{d}}\right)^{\alpha}}, \alpha\right) + \lambda_{\mathrm{u,s}}^{\mathrm{T}}\mathcal{Z}_{0, \iota_{OE}}^{\pi, \iota_{\mathrm{s}} + r_{\mathrm{d}}}\left(s; Q_{\mathrm{s}}\right), (2.47)$$

$$\mathcal{I}_3\left(s; Q_{\mathrm{d}}\right) = \pi \lambda_{\mathrm{d}}\left(\iota_{\mathrm{d}} + r_{\mathrm{d}}\right)^2\delta\left(\frac{sQ_{\mathrm{d}}}{\left(\iota_{\mathrm{d}} + r_{\mathrm{d}}\right)^{\alpha}}, \alpha\right) + \lambda_{\mathrm{d}}\mathcal{Z}_{0, \iota_{OF}}^{\pi, \iota_{\mathrm{d}} + r_{\mathrm{d}}}\left(s; Q_{\mathrm{d}}\right). \ (2.48)$$

The variables in the above formulas are defined as

$$\mathcal{F} \triangleq \lambda_{\mathrm{s}}^{\mathrm{D}}\delta\left(\gamma_{\mathrm{s}}^{\mathrm{D}}, \alpha\right) + \lambda_{\mathrm{u,s}}^{\mathrm{T}}C\left(\alpha\right)\left(\frac{Q_{\mathrm{s}}}{P_{\mathrm{s}}}\gamma_{\mathrm{s}}^{\mathrm{D}}\right)^{\frac{2}{\alpha}} + \frac{q_{\mathrm{D,s}}\lambda_{\mathrm{s}}}{\mathcal{A}_{\mathrm{D,s}}},$$

$$\mathcal{G} \triangleq C\left(\alpha\right)\left(\gamma_{\mathsf{s}}^{\mathsf{U}}\right)^{\frac{2}{\alpha}}\left(\lambda_{\mathsf{s}}^{\mathsf{D}}\left(\frac{P_{\mathsf{s}}}{Q_{\mathsf{s}}}\right)^{\frac{2}{\alpha}}+\lambda_{\mathsf{u},\mathsf{s}}^{\mathsf{T}}\right)+\frac{\left(1-q_{\mathsf{D},\mathsf{s}}\right)\lambda_{\mathsf{s}}}{\mathcal{A}_{\mathsf{U},\mathsf{s}}},$$

$$\mathcal{L}_{I_{\mathsf{d}\to\mathsf{D}}}(s\mid r\le\iota_{\mathsf{s}}) \;=\; \mathcal{L}_{I_{\mathsf{d}\to\mathsf{U}}}(s\mid r\le\iota_{\mathsf{s}})=\mathcal{L}_{I_{out}}(s\mid r)\exp\!\left(-\lambda_{\mathsf{d}}\mathcal{Z}_{0,l_{OA}}^{\pi,\iota_{\mathsf{s}}+r}(s;Q_{\mathsf{d}})\right),$$

$$\mathcal{L}_{I_{\mathsf{d}\to\mathsf{D}}}(s\mid r>\iota_{\mathsf{s}}) \;=\; \mathcal{L}_{I_{\mathsf{d}\to\mathsf{U}}}(s\mid r>\iota_{\mathsf{s}})=\mathcal{L}_{I_{out}}(s\mid r)\exp\!\left(-\lambda_{\mathsf{d}}\big(\mathcal{Z}_{0,0}^{\Theta,l_{OD}}(s;Q_{\mathsf{d}})\right.$$
$$\left.+\mathcal{Z}_{0,l_{OC}}^{\Theta,\iota_{\mathsf{s}}+r}(s;Q_{\mathsf{d}})+\mathcal{Z}_{\Theta,0}^{\pi,\iota_{\mathsf{s}}+r}(s;Q_{\mathsf{d}})\big)\right),$$

$$\mathcal{L}_{I_{out}}(s\mid r)=\exp\!\left(-\pi\lambda_{\mathsf{d}}\left(\iota_{\mathsf{s}}+r\right)^{2}\delta\!\left(\frac{sQ_{\mathsf{d}}}{\left(\iota_{\mathsf{s}}+r\right)^{\alpha}},\alpha\right)\right),\quad C\left(\alpha\right)=\frac{2\pi/\alpha}{\sin\left(2\pi/\alpha\right)},$$

$$\mathcal{Z}_{\theta_{l},\kappa_{l}}^{\theta_{u},\kappa_{u}}(s;Q)=(sQ)^{\frac{2}{\alpha}}\int_{\theta_{l}}^{\theta_{u}}\int_{\frac{\kappa_{l}^{2}}{(sQ)^{\frac{2}{\alpha}}}}^{\frac{\kappa_{u}^{2}}{(sQ)^{\frac{2}{\alpha}}}}\frac{1}{1+u^{\frac{\alpha}{2}}}\,du\,d\theta,\quad \delta\left(\beta,\alpha\right)=\int_{\beta^{-\frac{2}{\alpha}}}^{\infty}\frac{\beta^{\frac{2}{\alpha}}}{1+u^{\frac{\alpha}{2}}}\,du,$$

$$l_{OA}=\sqrt{\iota_{\mathsf{s}}^{2}-\left(r\sin\theta\right)^{2}}+r\cos\theta,\quad l_{OC}=\sqrt{\iota_{\mathsf{s}}^{2}-\left(r\sin\theta\right)^{2}}-r\cos\theta,$$

$$\Theta=\arcsin\!\left(\frac{\iota_{\mathsf{s}}}{r}\right),\quad l_{OD}=r\cos\theta-\sqrt{\iota_{\mathsf{s}}^{2}-\left(r\sin\theta\right)^{2}},$$

$$l_{OE}=\sqrt{\iota_{\mathsf{s}}^{2}-\left(r_{\mathsf{d}}\sin\theta\right)^{2}}+r_{\mathsf{d}}\cos\theta,\quad l_{OF}=\sqrt{\iota_{\mathsf{d}}^{2}-\left(r_{\mathsf{d}}\sin\theta\right)^{2}}+r_{\mathsf{d}}\cos\theta.$$

In the above equations, $\mathcal{F}$ and $\mathcal{G}$, respectively, correspond to the interference inflicted by DL SAPs and transmitting mobile users on the typical small cell receiver in DL and UL. $\mathcal{L}_{I_{\mathsf{d}\to\mathsf{D}}}(s\mid r\le\iota_{\mathsf{s}})$, $\mathcal{L}_{I_{\mathsf{d}\to\mathsf{D}}}(s\mid r>\iota_{\mathsf{s}})$ and $\mathcal{L}_{I_{\mathsf{d}\to\mathsf{U}}}(s\mid r\le\iota_{\mathsf{s}})$, $\mathcal{L}_{I_{\mathsf{d}\to\mathsf{U}}}(s\mid r>\iota_{\mathsf{s}})$ are the Laplace transforms of interference incurred by the active D2D transmitters on the typical small cell receiver in DL and UL, differentiated by the amplitude of the exclusion distance ι_{s}. $\mathcal{I}_{1}(s;P_{\mathsf{s}})$, $\mathcal{I}_{2}(s;Q_{\mathsf{s}})$ and $\mathcal{I}_{3}(s;Q_{\mathsf{d}})$ correspond to the interference incurred by the DL SAPs, transmitting mobile users and active D2D transmitters, respectively. The adoption of the CSMA scheme in our analysis leads to elaborate expressions of the coverage probability for the small cell tier. For the most general case, it involves triple integrals that can be efficiently solved by employing standard mathematical software packages. In the following section, we present the asymptotic analysis related to the protection threshold ρ_{s}, which simplifies the analysis substantially.

2.2.5.1   No D2D Transmissions

When $\rho_{\mathsf{s}}\to0$, we have $\lambda_{\mathsf{d}}\to0$. The coverage probability of small cell tier in DL and UL is, respectively [48],

$$\lim_{\rho_{\mathsf{s}}\to0}\mathbb{P}_{\mathsf{s}}^{\mathsf{D}}=\frac{q_{\mathsf{D},\mathsf{s}}\lambda_{\mathsf{s}}}{\lambda_{\mathsf{s}}^{\mathsf{D}}\mathcal{A}_{\mathsf{D},\mathsf{s}}\delta\left(\gamma_{\mathsf{s}}^{\mathsf{D}},\alpha\right)+\lambda_{\mathsf{u},\mathsf{s}}^{\mathsf{T}}\mathcal{A}_{\mathsf{D},\mathsf{s}}C\left(\alpha\right)\left(\frac{Q_{\mathsf{s}}}{P_{\mathsf{s}}}\gamma_{\mathsf{s}}^{\mathsf{D}}\right)^{\frac{2}{\alpha}}+q_{\mathsf{D},\mathsf{s}}\lambda_{\mathsf{s}}} \tag{2.49}$$

$$\lim_{\rho_{\mathsf{s}}\to0}\mathbb{P}_{\mathsf{s}}^{\mathsf{U}}=\frac{\left(1-q_{\mathsf{D},\mathsf{s}}\right)\lambda_{\mathsf{s}}}{C\left(\alpha\right)\left(\gamma_{\mathsf{s}}^{\mathsf{U}}\right)^{\frac{2}{\alpha}}\mathcal{A}_{\mathsf{U},\mathsf{s}}\left(\lambda_{\mathsf{s}}^{\mathsf{D}}\left(\frac{P_{\mathsf{s}}}{Q_{\mathsf{s}}}\right)^{\frac{2}{\alpha}}+\lambda_{\mathsf{u},\mathsf{s}}^{\mathsf{T}}\right)+\left(1-q_{\mathsf{D},\mathsf{s}}\right)\lambda_{\mathsf{s}}} \tag{2.50}$$

2.2.5.2　No Sensing for Small Cell Transmissions

When $\rho_{\mathsf{s}} \to \infty$, the active D2D transmitters form an MHP with the retaining probability $\beta = \frac{1-\exp(-\zeta\lambda_{\mathsf{u}}\mathcal{K}_{o,\mathsf{d}})}{\zeta\lambda_{\mathsf{u}}\mathcal{K}_{o,\mathsf{d}}}$

$\overset{(a)}{\approx} \frac{1}{\zeta\lambda_{\mathsf{u}}\mathcal{K}_{o,\mathsf{d}}}$, where (a) comes from $\zeta\lambda_{\mathsf{u}}\mathcal{K}_{o,\mathsf{d}} \gg 1$. Thereby the density of active

D2D transmitters is given by $\lambda_{\mathsf{d}} = \beta\zeta\lambda_{\mathsf{u}} \approx \frac{1}{\mathcal{K}_{o,\mathsf{d}}} = \frac{\alpha\left(\frac{\rho_{\mathsf{d}}}{Q_{\mathsf{d}}}\right)^{\frac{2}{\alpha}}}{2\pi\Gamma\left(\frac{2}{\alpha}\right)}$. The coverage

probability of small cell tier in DL and UL is, respectively, [48]

$$\lim_{\rho_{\mathsf{s}}\to\infty} \mathbb{P}_{\mathsf{s}}^{\mathsf{D}} \approx \frac{q_{\mathsf{D},\mathsf{s}}\lambda_{\mathsf{s}}}{\lambda_{\mathsf{s}}^{\mathsf{D}}\mathcal{A}_{\mathsf{D},\mathsf{s}}\delta\left(\gamma_{\mathsf{s}}^{\mathsf{D}},\alpha\right) + C\left(\alpha\right)\left(\gamma_{\mathsf{s}}^{\mathsf{D}}\right)^{\frac{2}{\alpha}}\mathcal{A}_{\mathsf{D},\mathsf{s}}\left(\lambda_{\mathsf{u},\mathsf{s}}^{\mathsf{T}}\left(\frac{Q_{\mathsf{s}}}{P_{\mathsf{s}}}\right)^{\frac{2}{\alpha}} + \frac{\alpha\left(\frac{\rho_{\mathsf{d}}}{P_{\mathsf{s}}}\right)^{\frac{2}{\alpha}}}{2\pi\Gamma\left(\frac{2}{\alpha}\right)}\right) + q_{\mathsf{D},\mathsf{s}}\lambda_{\mathsf{s}}},$$

$$\tag{2.51}$$

$$\lim_{\rho_{\mathsf{s}}\to\infty} \mathbb{P}_{\mathsf{s}}^{\mathsf{U}} \approx \frac{\left(1 - q_{\mathsf{D},\mathsf{s}}\right)\lambda_{\mathsf{s}}}{C\left(\alpha\right)\left(\gamma_{\mathsf{s}}^{\mathsf{U}}\right)^{\frac{2}{\alpha}}\mathcal{A}_{\mathsf{U},\mathsf{s}}\left(\lambda_{\mathsf{s}}^{\mathsf{D}}\left(\frac{P_{\mathsf{s}}}{Q_{\mathsf{s}}}\right)^{\frac{2}{\alpha}} + \lambda_{\mathsf{u},\mathsf{s}}^{\mathsf{T}} + \frac{\alpha\left(\frac{\rho_{\mathsf{d}}}{Q_{\mathsf{s}}}\right)^{\frac{2}{\alpha}}}{2\pi\Gamma\left(\frac{2}{\alpha}\right)}\right) + \left(1 - q_{\mathsf{D},\mathsf{s}}\right)\lambda_{\mathsf{s}}}.$$

$$\tag{2.52}$$

2.2.6　Network Throughput of D2D Enhanced HCN

With the coverage probability derive above, we derive the sum throughput of the two-tier network, where the bandwidth of each tier is normalized by W. We consider outage capacity with constant bit-rate coding, such that the total network throughput in DL and UL mode can be written as [48]

$$\mathcal{T}_{\mathsf{D}}\left(\eta;\rho_{\mathsf{s}};\rho_{\mathsf{d}}\right) = \eta\mathcal{T}_{\mathsf{m}}^{\mathsf{D}} + \left(1-\eta\right)\left(\mathcal{T}_{\mathsf{s}}^{\mathsf{D}} + \frac{1}{2}\mathcal{T}_{\mathsf{d}}\right), \tag{2.53}$$

$$\mathcal{T}_{\mathsf{U}}\left(\eta;\rho_{\mathsf{s}};\rho_{\mathsf{d}}\right) = \eta\mathcal{T}_{\mathsf{m}}^{\mathsf{U}} + \left(1-\eta\right)\left(\mathcal{T}_{\mathsf{s}}^{\mathsf{U}} + \frac{1}{2}\mathcal{T}_{\mathsf{d}}\right) \tag{2.54}$$

where $\mathcal{T}_i^{\mathsf{D}} = \lambda_i^{\mathsf{D}}\mathbb{P}_i^{\mathsf{D}}\log_2(1+\gamma_i^{\mathsf{D}})$, $\mathcal{T}_i^{\mathsf{U}} = \lambda_{\mathsf{u},i}^{\mathsf{T}}\mathbb{P}_i^{\mathsf{U}}\log_2(1+\gamma_i^{\mathsf{U}})$ and $\mathcal{T}_{\mathsf{d}} = \lambda_{\mathsf{d}}\mathbb{P}_{\mathsf{d}}\log_2(1+\gamma_{\mathsf{d}})$. Half of the D2D outage capacity is included in the DL and UL network throughput, respectively. Note that in (2.53) and (2.54), the load of base stations is incorporated in the calculation of *active* transmitter density λ_i^{D} and $\lambda_{\mathsf{u},i}^{\mathsf{T}}$ with the empty cells being excluded. Note that in (2.53) and (2.54), we have normalized the bandwidth and the network coverage area. Thus, the derived network throughput is the data rate (in unit bps) per Hz per m^2.

2.2.7　Validation of Network Model

To validate approximations considered in this chapter, we do extensive simulations. All simulations are performed over a square window of $5{,}000 \times 5{,}000$ m^2 with $10{,}000$ iterations. Unless otherwise specified, we use the following default values of the system parameters: $\lambda_{\mathsf{m}} = 1/(\pi500^2)m^{-2}$, $\alpha = 4$,

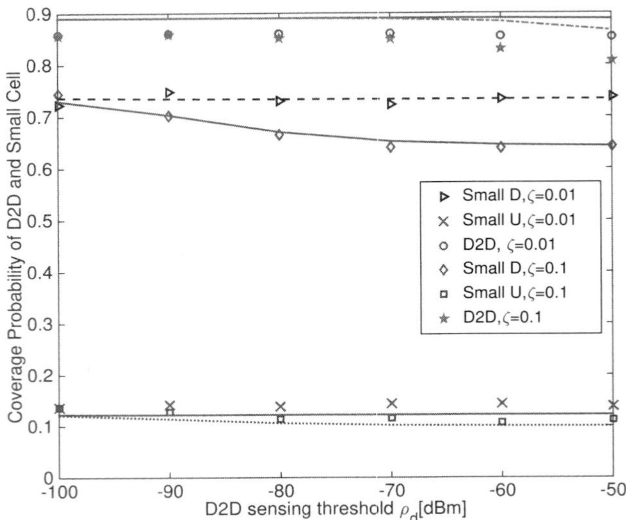

Figure 2.6: Comparison of D2D and small cell tier coverage probability from simulation (markers) and theoretical analysis (lines) as a function of ρ_s, for $\lambda_\mathrm{s} = 5\lambda_\mathrm{m}$, $\lambda_\mathrm{u} = 100\lambda_\mathrm{m}$, $\{q_{\mathrm{D,m}}, q_{\mathrm{D,s}}\} = \{0.5, 0.5\}$, $\rho_\mathrm{s} = -60$ dBm, and $\{B_{\mathrm{D,m}}, B_{\mathrm{D,s}}, B_{\mathrm{U,m}}, B_{\mathrm{U,s}}\} = \{1, 1, 1, 1\}$.

$P_\mathrm{m} = 46$ dBm, $Q_\mathrm{m} = 20$ dBm, $P_\mathrm{s} = 26$ dBm, $Q_\mathrm{s} = 10$ dBm, $Q_\mathrm{d} = 0$ dBm, $\gamma_\mathrm{m}^\mathrm{D} = \gamma_\mathrm{m}^\mathrm{U} = \gamma_\mathrm{s}^\mathrm{D} = \gamma_\mathrm{s}^\mathrm{U} = 0$ dB, $\gamma_\mathrm{d} = 0$ dB, $r_\mathrm{d} = 20$ m, $\rho_\mathrm{s} = \rho_\mathrm{d} = -60$ dBm, $\mu = 0.5$, and $\epsilon = 10^{-5}$. For the network throughput, we set the network bandwidth $W = 10$ MHz, and the macrocell coverage area is S $= \pi 500^2$ m^2. Note that the approximation is caused by the following factors: (1) modeling the combined effect of PHP and MHP with an independent thinning of a PPP, (2) neglecting the coupling between locations of mobile users and base stations in the UL transmission, (3) replacing the instantaneous exclusion distance by ι_s and ι_d constrained by a small misdetection threshold ϵ.

Figure 2.6 indicates that the PPP approximation is accurate for low and high values of ρ_d. Low values of ρ_d correspond to large exclusion distances ι_s, leading to a low retaining probability β. The good agreement between simulation and analysis can be explained by the fact that the smaller density of D2D transmitters leads to little interference. For high values of ρ_d and corresponding small exclusion distances ι_d, the density of the active D2D transmitters approaches that of the initial PPP, which eliminates the inaccuracy caused by the approximation. The middle range of values of ρ_d results in inaccuracy on the coverage probability. Figure 2.6 also depicts that both the coverage probabilities of small cell tier and D2D user deteriorate with ρ_d. Similar effect can be seen for ρ_s. This is due to the fact that the retaining probability and corresponding λ_d increase with ρ_s and ρ_d. Thus, in terms of the overall coverage probability for infrastruc-

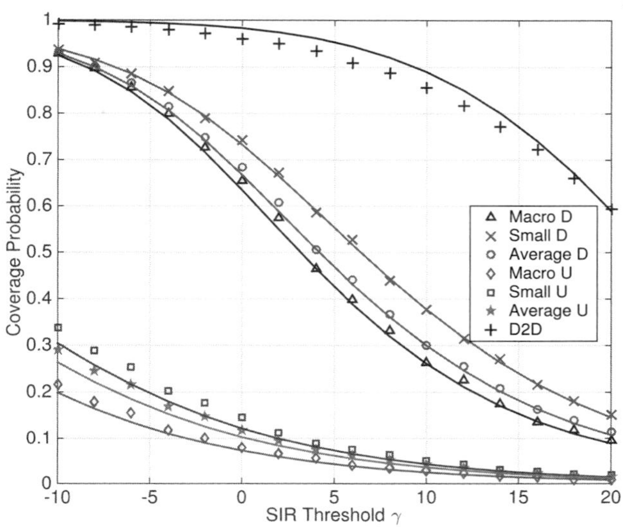

Figure 2.7: Comparison of coverage probability from simulation (markers) and theoretical analysis (lines) as a function of SIR threshold, ($\lambda_s = 5\lambda_m$, $\lambda_u = 100\lambda_m$, $\{q_{D,m}, q_{D,s}\} = \{0.5, 0.5\}$, $\rho_s = \rho_d = -60$ dBm, and $\{B_{D,m}, B_{D,s}, B_{U,m}, B_{U,s}\} = \{1, 1, 1, 1\}$).

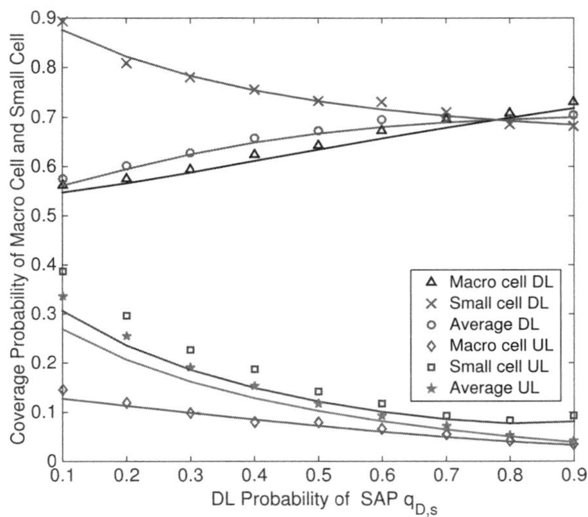

Figure 2.8: Downlink coverage probability versus λ_s, (a) for a macrotier, small cell tier, and D2D user, and (b) for overall coverage of the infrastructure, ($\lambda_m = \frac{1}{\pi 500^2}$, $\lambda_u = 10^3\lambda_m$, $q_{D,m} = 0.5$, $\{B_{D,m}, B_{D,s}, B_{U,m}, B_{U,s}\} = \{1, 1, 1, 1\}$, $\rho_s = \rho_d = -60$ dBm).

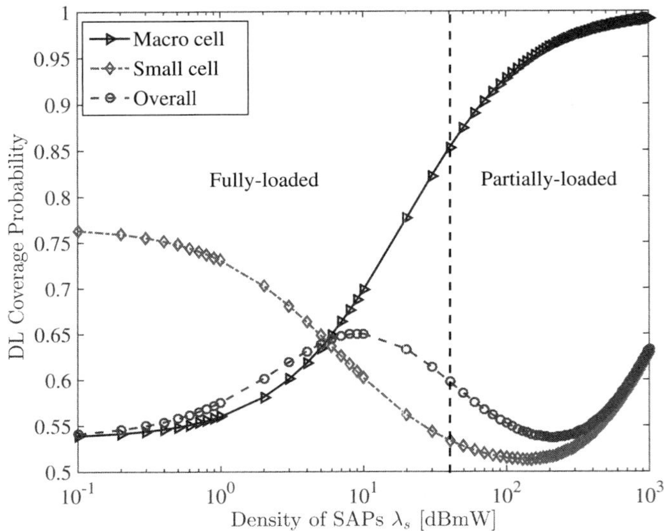

Figure 2.9: Downlink coverage probability versus λ_s, for $\lambda_m = \frac{1}{\pi 500^2}$, $\lambda_u = 10^3 \lambda_m$, $\{q_{D,m}, q_{D,s}\} = \{0.5, 0.5\}$, $\{B_{D,m}, B_{D,s}, B_{U,m}, B_{U,s}\} = \{1, 1, 1, 1\}$, and $\rho_s = \rho_d = -60$ dBm).

ture based transmissions and typical D2D user, the optimal sensing threshold is given by $\rho_s^\star = 0$ and $\rho_d^\star = 0$. However, the absence of D2D transmissions results in reduced network throughput.

Figure 2.7 represents the coverage probability as a function of SIR threshold. It shows that the coverage probabilities of macrocell, small cell, and D2D users decrease with the increase of SIR threshold. DL coverage probability is larger than UL coverage probability due to the fact that DL transmit power is higher than UL transmit power, while D2D coverage is the largest, which can be explained by the short range of the D2D link.

2.2.8 Numerical Analysis and Discussion

In this section, we evaluate how the important network parameters, such as base station density and bias factor, affect the load-aware coverage probability.

As shown in Figure 2.8, $\mathbb{P}_i^U$ is a decreasing function of $q_{D,i}$ due to the large difference between DL transmit power and UL transmit power. How the UL/DL configuration $q_{D,i}$ affects $\mathbb{P}_i^D$ is not very explicit, because increasing $q_{D,i}$ gives rise to a reduction of the UL interference and a surge of the DL interference. However, for a given set of system parameters, we find for each tier i that the relative transmit power $\frac{Q_i}{P_i}$ determines whether $\mathbb{P}_i^D$ is dominated by DL or UL interference. For instance, Figure 2.8 shows that in the current parameter setting, $\mathbb{P}_s^D$ decreases with the increase of $q_{D,s}$.

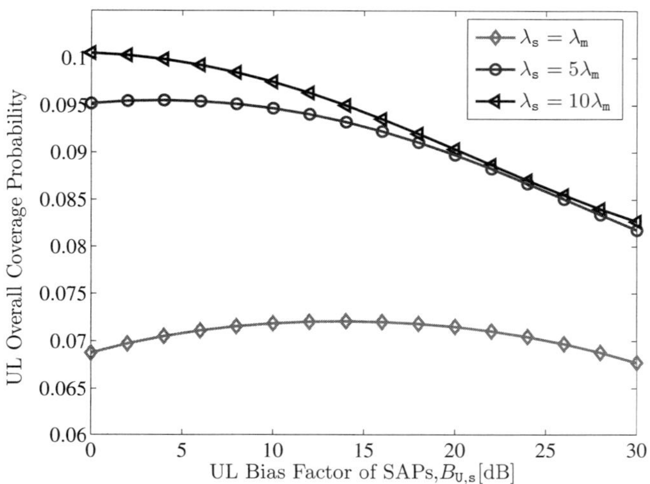

Figure 2.10: Overall uplink coverage probability as a function of UL bias factor of SAPs $B_{\mathrm{U,s}}$, ($\lambda_{\mathrm{u}} = 10^3 \lambda_{\mathrm{m}}$, $\{q_{\mathrm{D,m}}, q_{\mathrm{D,s}}\} = \{0.5, 0.5\}$, $\{B_{\mathrm{D,m}}, B_{\mathrm{U,m}}, B_{\mathrm{D,s}}\} = \{1, 1, 1\}$, and $\zeta = 0.01$, $\rho_{\mathrm{s}} = \rho_{\mathrm{d}} = -60$ dBm).

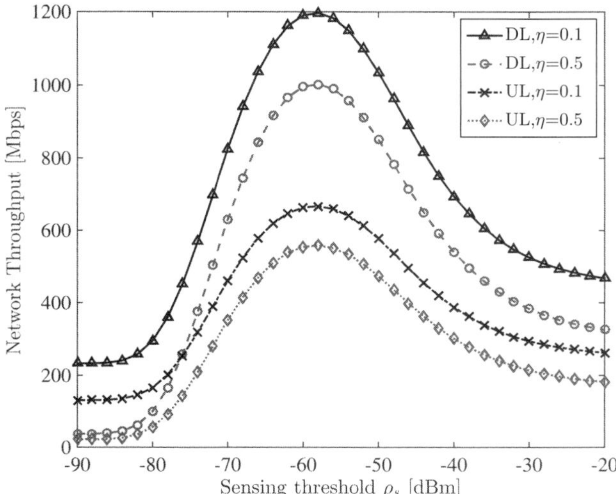

Figure 2.11: Total network throughput versus protection threshold ρ_{s}, with $\rho_{\mathrm{d}} = -20$ dBm, ($\lambda_{\mathrm{u}} = 10^4 \lambda_{\mathrm{m}}$, $\lambda_{\mathrm{s}} = 100 \lambda_{\mathrm{m}}$, $\{q_{\mathrm{D,m}}, q_{\mathrm{D,s}}\} = \{0.5, 0.5\}$, $\zeta = 0.05$).

In Figure 2.9, we evaluate the variation of the DL load-aware coverage probability $\mathbb{P}_{\mathrm{m}}^{\mathrm{D}}$, $\mathbb{P}_{s}^{\mathrm{D}}$, and $\bar{\mathbb{P}}_{\mathrm{D}}$ as a function of λ_{s}. In terms of traffic load, the network evolves from a fully loaded sparse network to a partially loaded dense network. We observe that $\mathbb{P}_{\mathrm{m}}^{\mathrm{D}}$ increases monotonously with λ_{s}, which can be ascribed to the handover of macro mobile users with low SIR to the small cell tier and the corresponding reduction of interference in the macrotier. With respect to the small cell tier, the small cell network interference increases with λ_{s}, while the activity of D2D users diminishes exponentially with λ_{s} as can be verified in (2.35). These opposite effects are reflected in the load-aware coverage probability for the small cell tier. As λ_{s} increases, the overall coverage probability $\bar{\mathbb{P}}_{\mathrm{D}}$ first increases, then decreases, and increases again. It indicates that an optimal λ_{s} can be found in the feasible region of small cell densities. As λ_{s} increases, the network load moves into the lightly loaded regime, where the aggregate small cell interference is constrained by the density of mobile users. As $\lambda_{\mathrm{s}} \to \infty$, we have $\mathcal{A}_{\mathrm{D,s}} \to 1$ and $\bar{\mathbb{P}}_{\mathrm{D}} \to 1$. As opposed to the fully loaded traffic model with constant coverage probability in the asymptotic regime [47, 51], this result highlights the usefulness of the load-aware model to capture coverage probability in realistic conditions. Given a good estimate of the user density, the proposed analytical framework allows us to find the small cell density within the realistic regime that optimizes overall coverage probability.

To see the effect of UL bias factor on network performance, we depict overall UL coverage probability as a function of the UL bias factor of small cell tier $B_{\mathrm{U,s}}$ in Figure 2.11 as follows. We observe that increasing the density of SAPs λ_{s} decreases the optimal $B_{\mathrm{U,s}}$. This is due to the fact that a larger λ_{s} attracts transmitting mobile users to associate with the small cell tier, which inflicts more interference on the UL SAPs. Decreasing $B_{\mathrm{U,s}}$ helps to increase overall coverage probability by shifting small cell mobile users with low SIR to the macrotier.

Figure 2.11 depicts the total network throughput as a function of ρ_{s}. We observe that the network throughput exhibits a concave behavior with respect to ρ_{s}. This is caused by the opposite effects of ρ_{s} on λ_{d} and the coverage probability of the small cell tier and typical D2D user, a trade-off between coverage probability and D2D user activity that is made explicit in the expressions of the network throughput (2.53) and (2.54). Similar results can be derived for the effect of ρ_{d} on the network throughput. Figure 2.11 also shows that in the dense scenario ($\lambda_{\mathrm{s}} = 100\lambda_{\mathrm{m}}$), giving more bandwidth to the small cell tier can increase the total network throughput. The presented results show that the proposed analytical framework can be used to determine $\rho_{\mathrm{s}}^{\star}$ or $\rho_{\mathrm{d}}^{\star}$ that maximize the total network throughput.

2.2.9   Conclusion

In this chapter, we studied a two-tier D2D enhanced HCN operating with dynamic TDD, where the D2D transmitters follow a CSMA scheme. We proposed a simple PPP model for the active D2D users and verified the accuracy by extensive simulations. We presented an analytical framework to evaluate the load-aware coverage probability and network throughput. The proposed model allows us to analyze the nontrivial system behavior of dynamic TDD networks and to quantify the effect of most important network parameters such as the UL/DL configuration, base station density, and bias factor on coverage probability, and the bandwidth partition on total network throughput. We provided guidelines on the optimal design of the network access scheme. Possible future directions to extend this work are to include a dynamic traffic model in our framework and consider the spatiotemporal correlations in the dynamic TDD network.

REFERENCES

[1] K. Mallinson, "The 2020 vision for LTE," http://www.fiercewireless.com/europe/story/mallinson-2020-vision-lte/2012-06-20, Tech. Rep.

[2] Y. Lin and Y. Hsu, "Multihop cellular: a new architecture for wireless communications," in *Proc. IEEE INFOCOM*, Tel Aviv, Israel, Mar. 2000.

[3] P. Li, X. Huang, and Y. Fang, "Capacity scaling of multihop cellular networks," in *Proc. IEEE INFOCOM*, Shanghai, China, Apr. 2011.

[4] H. Luo, R. Ramjee, P. Sinha, L. Li, and S. Lu, "Ucan: a unified cellular and ad-hoc network architecture," in *Proc. ACM MobiCom*, ser. MobiCom '03. New York, NY, USA: ACM, June 2003.

[5] A.Ghosh, N.Mangalvedhe, R.Ratasuk, B.Mondal, M.Cudak, E.Visotsky, T.A.Thomas, J.G.Andrews, P.Xia, H.S.Jo, H.S.Dhillon, and T.D.Novlan, "Heterogeneous cellular networks: From theory to practice," *IEEE Commun. Mag.*, vol. 50, no. 6, pp. 54–64, June 2012.

[6] M. Haenggi, J. Andrews, F. Baccelli, O. Dousse, and M. Franceschetti, "Stochastic geometry and random graphs for the analysis and design of wireless networks," *IEEE J. Sel. Areas Commun.*, vol. 27, no. 7, pp. 1029–1046, Sep. 2009.

[7] P. Gupta and P. Kumar, "The capacity of wireless networks," *IEEE Trans. Inf. Theory*, vol. 46, no. 2, pp. 388–404, Mar. 2000.

[8] F. Xue, L. Xie, and P. Kumar, "The transport capacity of wireless networks over fading channels," *IEEE Trans. Inf. Theory*, vol. 51, no. 3, pp. 834–847, Mar. 2005.

[9] M. Franceschetti, D. Migliore, and P. Minero, "The capacity of wireless networks: Information-theoretic and physical limits," *IEEE Trans. Inf. Theory*, vol. 55, no. 8, pp. 3413–3424, Aug. 2009.

[10] M.Grossglauser and D.N.C.Tse, "Mobility increases the capacity of ad hoc wireless networks," *IEEE/ACM Trans. Networking*, vol. 10, no. 4, pp. 477–486, 2002.

[11] S. P. Weber, X. Yang, J. Andrews, and G. De Veciana, "Transmission capacity of wireless ad hoc networks with outage constraints," *IEEE Trans. Inf. Theory*, vol. 51, no. 12, pp. 4091–4102, Dec. 2005.

[12] S. Weber, J. Andrews, and N. Jindal, "An overview of the transmission capacity of wireless networks," *IEEE Trans. Commun.*, vol. 58, no. 12, pp. 3593–3604, Dec. 2010.

[13] C. Jaeweon and Z. J. Haas, "On the throughput enhancement of the downstream channel in cellular radio networks through multihop relaying," *IEEE J. Sel. Areas Commun.*, vol. 22, no. 7, pp. 1206–1219, Sep. 2004.

[14] L. Law, K. Pelechrinis, S. Krishnamurthy, and M. Faloutsos, "Downlink capacity of hybrid cellular ad hoc networks," *IEEE/ACM Trans. Networking*, vol. 18, no. 1, pp. 243–256, Feb. 2010.

[15] K.-H. Lee, K.-Y. Han, J.-Y. Song, and D.-H. Cho, "Capacity enhancement of uplink channel through spatial reuse in multihop cellular networks," *IEEE Comms. Lett.*, vol. 10, no. 2, pp. 76–78, Feb. 2006.

[16] M. Hossain and A. Mammela, "Effect of user density and traffic volume on uplink capacity of multihop cellular network," in *Proc. IEEE ICWMC*, Cannes, La Bocca, Aug. 2009.

[17] H.S.Dhillon, R.K.Ganti, F.Baccelli, and J.G.Andrews, "Modeling and analysis of k-tier downlink heterogeneous cellular networks," *IEEE J. Sel. Areas Commun.*, vol. 30, no. 3, pp. 550–560, April 2012.

[18] S. Mukherjee, "Distribution of downlink sinr in heterogeneous cellular networks," *IEEE J. Sel. Areas Commun.*, vol. 30, no. 3, pp. 575–585, April 2012.

[19] F. Baccelli and B. Blaszczyszyn, *Stochastic Geometry and Wireless Networks, Volume 1: Theory.* NOW, Dec. 2009.

[20] R.K.Ganti and M.Haenggi, "Spatial analysis of opportunistic downlink relaying in a two-hop cellular system," *IEEE Trans. Commun.*, vol. 60, no. 5, pp. 1443–1450, 2012.

[21] P. H. J.Nardelli, P.Cardieri, and M.Latva-aho, "Efficiency of wireless networks under different hopping strategies," *IEEE Trans. Wireless Commun.*, vol. 11, no. 1, pp. 15–20, Jan. 2012.

[22] M. Haenggi, "On distances in uniformly random networks," *IEEE Trans. Inf. Theory*, vol. 51, no. 10, pp. 3584–3586, Oct. 2005.

[23] J. Wen, M. Sheng, X. Wang, J. Li, and H. Sun, "On the capacity of downlink multi-hop heterogeneous cellular networks," *IEEE Trans. Wireless Commun.*, vol. 13, no. 8, pp. 4092–4103, Aug. 2014.

[24] D. Feng, L. Lu, Y. Yuan-Wu, G. Ye Li, S. Li, and G. Feng, "Device-to-device communications in cellular networks," *IEEE Commun. Mag.*, vol. 52, no. 4, pp. 49–55, Apr. 2014.

[25] L. Song, D. Niyato, Z. Han, and E. Hossain, "Game-theoretic resource allocation methods for device-to-device communication," *IEEE Wireless Commun.*, vol. 21, no. 3, pp. 136–144, June 2014.

[26] T. Q. S. Quek, G. de la Roche, I. Gven, and M. Kountouris, *Small Cell Networks: Deployment, PHY Techniques, and Resource Management.* New York, NY, USA: Cambridge Univ. Press, 2013.

[27] M. Wildemeersch, T. Q. S. Quek, C. Slump, and A. Rabbachin, "Cognitive small cell networks: Energy efficiency and trade-offs," *IEEE Trans. Commun.*, vol. 61, no. 9, pp. 4016–4029, Sept. 2013.

[28] W. K. D. Stoyan and J. Mecke, *Stochastic geometry and its applications.* Chichester: Wiley, 1995.

[29] X. Lin, J. Andrews, and A. Ghosh, "Spectrum sharing for device-to-device communication in cellular networks," *IEEE Trans. Wireless Commun.*, vol. 13, no. 12, pp. 6727–6740, Dec. 2014.

[30] H. ElSawy, E. Hossain, and M.-S. Alouini, "Analytical modeling of mode selection and power control for underlay D2D communication in cellular networks," *IEEE Trans. Commun.*, vol. 62, no. 11, pp. 4147–4161, Nov 2014.

[31] Q. Ye, M. Al-Shalash, C. Caramanis, and J. Andrews, "Resource optimization in device-to-device cellular systems using time-frequency hopping," *IEEE Trans. Wireless Commun.*, vol. 13, no. 10, pp. 5467–5480, Oct 2014.

[32] M. Sheng, J. Liu, X. Wang, Y. Zhang, H. Sun, and J. Li, "On transmission capacity region of D2D integrated cellular networks with interference management," *accepted by IEEE Trans. Commun.*, 2015.

[33] C. Xu, L. Song, Z. Han, Q. Zhao, X. Wang, X. Cheng, and B. Jiao, "Efficiency resource allocation for device-to-device underlay communication systems: A reverse iterative combinatorial auction based approach," *IEEE J. Sel. Areas Commun.*, vol. 31, no. 9, pp. 348–358, Sept. 2013.

[34] Y. Li, D. Jin, J. Yuan, and Z. Han, "Coalitional Games for Resource Allocation in the Device-to-Device Uplink Underlaying Cellular Networks," *IEEE Trans. Wireless Commun.*, vol. 13, no. 7, pp. 3965–3977, July 2014.

[35] S. Sesia, I. Toufik, and M. Baker, *LTE- The UMTS Long Term Evolution: From Theory to Practice.* New Jersey, NJ, USA: John Wiley & Sons, 2011.

[36] Z. Shen, A. Khoryaev, E. Eriksson, and X. Pan, "Dynamic uplink-downlink configuration and interference management in TD-LTE," *IEEE Commun. Mag.*, vol. 50, no. 11, pp. 51–59, Nov. 2012.

[37] M. S. ElBamby, M. Bennis, W. Saad, and M. Latva-aho, "Dynamic Uplink-Downlink Optimization in TDD-based Small Cell Networks," in *Proc. IEEE INFOCOM Workshops*, Toronto, Canada, Apr. 27 - May 2, 2014, pp. 712–717.

[38] J. Li, S. Farahvash, M. Kavehrad, and R. Valenzuela, "Dynamic TDD and fixed cellular networks," *IEEE Commun. Lett.*, vol. 4, no. 7, pp. 218–220, Jul. 2000.

[39] V. Chandrasekhar, J. G. Andrews, T. Muharemovict, Z. Shen, and A. Gatherer, "Power control in two-tier femtocell networks," *IEEE Trans. Wireless Commun.*, vol. 8, no. 8, pp. 4316–4328, Aug. 2009.

[40] M. Wildemeersch, T. Q. S. Quek, M. Kountouris, A. Rabbachin, and C. Slump, "Successive interference cancellation in heterogeneous networks," *IEEE Trans. Commun.*, vol. 62, no. 12, pp. 4440–4453, Dec. 2014.

[41] V. Chandrasekhar and J. G. Andrews, "Spectrum allocation in tiered cellular networks," *IEEE Trans. Wireless Commun.*, vol. 57, no. 10, pp. 3059–3068, Oct. 2009.

[42] M. Haenggi, "Mean interference in hard-core wireless networks," *IEEE Commun. Lett.*, vol. 15, no. 8, pp. 792–794, Aug. 2011.

[43] H. ElSawy and E. Hossain, "A Modified Hard Core Point Process for Analysis of Random CSMA Wireless Networks in General Fading Environments," *IEEE Trans. Commun.*, vol. 61, no. 4, pp. 1520–1534, Apr. 2013.

[44] C. han Lee and M. Haenggi, "Interference and outage in poisson cognitive networks," *IEEE Trans. Wireless Commun.*, vol. 11, no. 4, pp. 1392–1401, Apr. 2012.

[45] T. V. Nguyen and F. Baccelli, "A stochastic geometry model for cognitive radio networks," *The Computer Journal*, vol. 55, no. 5, pp. 534 – 552, Jul. 2012.

[46] T. Novlan, H. Dhillon, and J. Andrews, "Analytical Modeling of Uplink Cellular Networks," *IEEE Trans. Wireless Commun.*, vol. 12, no. 6, pp. 2669–2679, June 2013.

[47] J. G. Andrews, F. Baccelli, and R. Ganti, "A Tractable Approach to Coverage and Rate in Cellular Networks," *IEEE Trans. Commun.*, vol. 59, no. 11, pp. 3122–3134, Nov. 2011.

[48] H. Sun, M. Wildemeersch, M. Sheng, and T. Q. S. Quek, "D2D Enhanced Heterogeneous Cellular Networks With Dynamic TDD," *IEEE Trans. Wireless Commun.*, vol. 14, no. 8, pp. 4204–4218, Aug 2015.

[49] S. M. Yu and S.-L. Kim, "Downlink capacity and base station density in cellular networks," in *Proc. IEEE WiOpt*, Tsukuba Science City, Japan, May 13-17, 2013, pp. 119–124.

[50] Y. S. Soh, T. Q. S. Quek, M. Kountouris, and G. Caire, "Cognitive hybrid division duplex for two-tier femtocell networks," *IEEE Trans. Wireless Commun.*, vol. 12, no. 10, pp. 4852–4865, Oct. 2013.

[51] H. Dhillon, R. Ganti, F. Baccelli, and J. Andrews, "Modeling and analysis of K-Tier downlink heterogeneous cellular networks," *IEEE J. Sel. Areas Commun.*, vol. 30, no. 3, pp. 550–560, 2012.

Chapter 3

Interference Management Framework and Design Issues

Future wireless networks are expected to be ultradense heterogeneous networks [1]. Various classes of low-power nodes (LPNs), such as micro base stations, pico eNBs, home eNBs (also called femtocells), relays and DAS (also called remote radio heads or RRHs) are deployed ultradensely throughout the macrocell network. Within these networks, there are different types of resources, including space resources, spectrum resources, storage resources and computation resources. A major factor that limits the achievable network performance (e.g., coverage probability and network capacity) is interference, which is caused by the conflict of resource allocation among different network equipment. Severe interference will occur if the resources are not carefully managed. Therefore, from this perspective, the nature of interference management is, in fact, the management of resources.

To tackle the difficulty in interference management in heterogeneous wireless networks, in this chapter, we first propose the concept of resource mobility, where the network resources can be flexibly allocated within the whole network and moved from lightly loaded cells to heavily loaded cells. Based on the resource mobility concept, a new interference management framework is proposed, where the necessary functional entities are introduced to support the resource mobility capability. The functions executed by all the entities is described in detail. Then, we focus on two important aspects in interference management: interference modeling and interference management techniques.

3.1 Interference Management Framework Based on Resource Mobility

In this section, we first introduce the concept of resource mobility, and then construct an interference management framework based on resource mobility.

(a) Before resource movement

(b) After resource movement

Figure 3.1: The illustration of resource mobility.

3.1.1   Resource Mobility

From the view of the whole network, user traffic is not uniformly distributed and is dynamically changing. Network resources are also nonuniformly distributed. Therefore, there is a challenge when attempting to match the uneven resources with the uneven traffic demand. Taking Figure 3.1 as an example, in those dark cells, due to the high traffic requirements, they have a heavy load and are short of resources, while in those light cells, they have a light load and have unused resources. Therefore, we can move the resources in those light cells to the dark cells. We call this type of resource movement as resource mobility. Specifically, resource mobility has the following three novel concepts.

3.1.1.1   Resource Migration Scenario

Resource migration is a direct method to overcome the mismatch between the nonuniformly distributed resources and traffic. Here, we will give an example of resource migration [2]. As shown in Figure 3.2, we consider the resource management in a three-tier HetNet, which consist of a wireless metropolitan area network (WMAN), a cellular network, and a wireless local area network (WLAN). A dynamic hierarchy resource management approach (DHRM) was

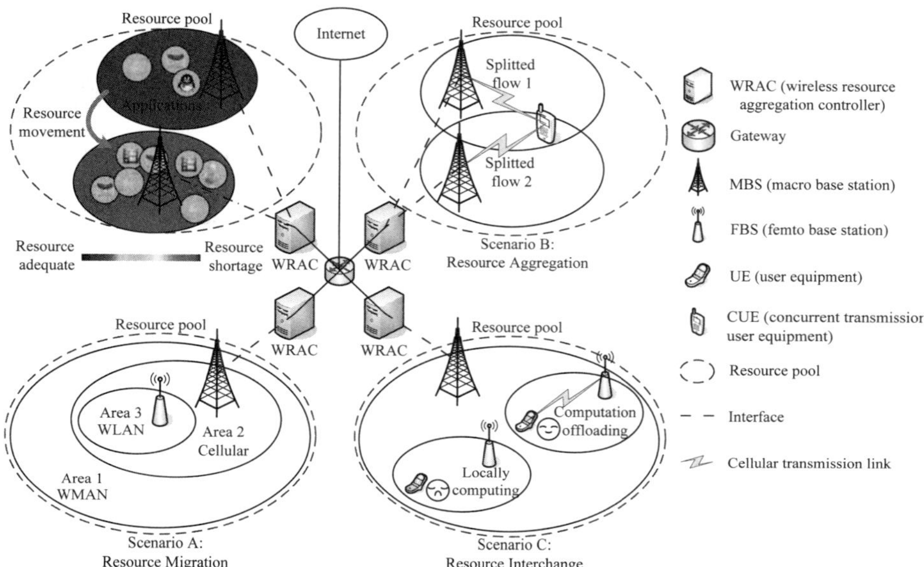

Figure 3.2: The network and resource management scenario.

proposed to migrate network resource with the change of network traffic and thus improve network capacity. In DHRM, the method of wavelet neural network and Wiener prediction are first adopted to obtain the variation of traffic distribution in a different time scale. Based on these predicted results, the available hierarchical resources in the networks are flexibly allocated such that the resources can match well with traffic. Further, we analyze the performance of a simplified DHRM in HetNets with heavy tailed traffic via multidimensional Markov model in [3]. Next, we show that resource migration improves network capacity via simulation. Figure 3.3 shows the number of accessed users in DHRM and internetwork static resource management (ISRM [2]) varying with arrival rate and time. From this figure, we find that DHRM outperforms ISRM about 20% on average in terms of the total number of accessed users. This is because the resources are dynamically managed according to the variation of network traffic in DHRM. In addition, with the increasing arrival rate of users, the total number of accessed users of DHRM and those of ISRM increase at first and then remain stable after some arrival rate. This is obvious since the resources in the network are limited.

3.1.1.2 Resource Aggregation Scenario

To overcome the mismatch between resources and the demand of the applications, discrete resources can be accumulated to provide guaranteed quality of service (QoS), which is defined as resource aggregation. What we should consider is which type of resource could be accumulated and which entity would

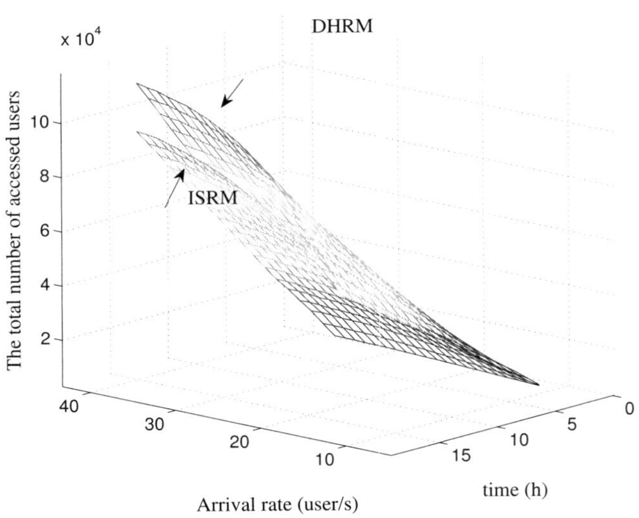

Figure 3.3: Total number of accessed users in a resource migration scenario.

execute resource aggregation. Moreover, in order to enhance system performance, how many resources are to be accumulated is also a problem well worth addressing. We give a simple example of resource aggregation for a large volume traffic transmission [4] in this section. In HetNets, the concurrent transmission user equipment (CUE) could obtain resource aggregation gain by simultaneously combining transmissions over several RATs. As shown in the resource aggregation scenario in Figure 3.1, we consider the concurrent transmission over two RATs (i.e., cellular networks and wireless LAN networks). The centric WRAC manages different RATs and executes flow-splitting strategy. When either cellular (RAT_1) or wireless LAN (RAT_2) cannot support a given traffic, which has a data rate requirement R_{req}, we could aggregate the cellular links and wireless LAN links to support the traffic. In other words, using resource aggregation, $R_1 + R_2 \geq R_{req}$ can be guaranteed, where R_1 and R_2 denote the rate carried by RAT_1 and RAT_2, respectively. Therefore, the requirement of CUE could be satisfied with a larger probability. Even if the requirement of CUE could be guaranteed by a single link, resource aggregation still has benefits in other aspects. For example, with the goal of maximizing the system EE, we design a splitting strategy maximum EE concurrent transmission (MECT) in [4]. Here, EE (energy efficiency) is defined as the aggregate bit rate that is achievable while consuming a given power. With MECT, higher EE could be obtained by aggregating two RATs to transmit concurrently when the data rate requirement is larger than R_b, which is derived in a simple formula as follows,

$$R_b = B_1 \log_2 \left(g_1/g_2 \right), \tag{3.1}$$

where B_1 is the available bandwidth in RAT_1. As well, g_1 and g_2 denote the channel power gains in RAT_1 and RAT_2, respectively. Furthermore, the optimal rate allocation of each RAT is given by (3.2).

$$\begin{cases} R_1^{\star} = \dfrac{B_1 R_{req} - B_1 B_2 \log_2\left(\frac{g_2}{g_1}\right)}{B_1 + B_2} \\ R_2^{\star} = \dfrac{B_2 R_{req} + B_1 B_2 \log_2\left(\frac{g_2}{g_1}\right)}{B_1 + B_2} \end{cases}, \qquad (3.2)$$

In (3.2), $R_1^{\star}$ and $R_2^{\star}$ are the optimal rate allocation of each RAT, and B_2 is the available bandwidth in RAT_2. Figure 3.4(a) shows the optimal rate of each RAT. RAT_1 carries all the traffic at the beginning. With an increasing data rate, traffic flow will be distributed into both RATs. Figure 3.4(b) shows EE performance of HetNets in MECT, Fixed Splitting (FS), and single RAT transmission strategy. With an increasing data rate, EE performance of these four strategies decreases. However, due to the resource aggregation in MECT, EE performance of MECT decreases much more slowly than that of a single RAT, which could be considered as the resource aggregation gain in improving EE. Typically, when the data rate requirement is bps, the gain brought by resource aggregation can reach about 300%. Additionally, MECT outperforms FS in terms of EE performance. This is because the spitting ratio in MECT could change with the variation of the data rate requirement. It is worth noting that RAT_1 will be used exclusively when data requirement is less than R_b, which is consistent with the theoretical analysis.

3.1.1.3 Resource Interchange Scenario

With resource interchange, different types of resources could be interchanged according to some rules. For instance, when users cannot support the computation-intensive applications, some communication resources can be utilized so that huge demand for computation resources could be alleviated. In order to implement resource interchange, some challenges should be tackled. First, we should know which resources can be interchanged. Moreover, with the purpose of achieving enhanced system performance, we should know how many resources should be interchanged into other type(s) of resources.

Next, we will give an example to explain resource interchange. With the proliferation of resource-hungry applications, such as speech recognition (e.g., Siri), online games, and reality augmentation, it is very challenging for UE to process these applications due to limited resources of energy, computation and storage. Even if these applications can be supported, it will largely reduce battery lifetime, thereby shortening the standby time. In [5], we propose a resource interchange (RI) enabled strategy to address this problem. Here, we focus on the femtocloud scenario as shown in the resource interchange scenario in Figure 3.1. When UE has an application to process, it should judiciously determine whether to offload computation as well as what portion should be offloaded. Once the offloading decision is made, three phases need to be performed sequentially. First, the UE sends data to the cloud through the uplink channel.

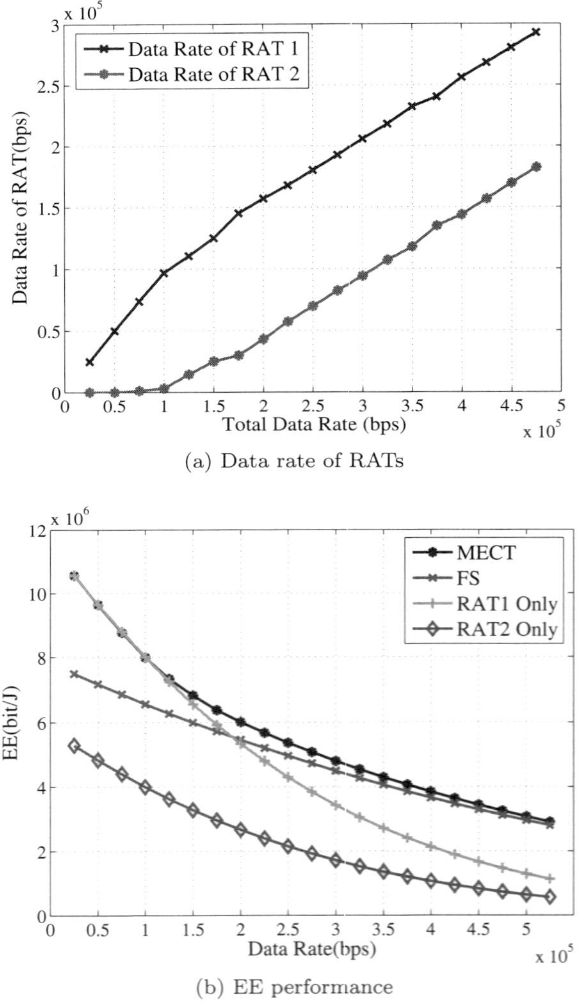

(a) Data rate of RATs

(b) EE performance

Figure 3.4: Optimal data rate and EE performance in a resource aggregation scenario.

Second, the cloud executes the computation task. Finally, the results are sent back to the UE through the downlink channel. In these procedures, UE employs the concept of resource interchange by utilizing communication resources to reduce the huge demand for computation resources. Through interchanging and jointly scheduling the computation and communication resources, the applications with the more stringent latency requirement can be supported and the standby time of UE can be prolonged. Furthermore, the optimal power used for the communication procedure and the optimal computing speed of UE are both derived in [5].

Figure 3.5(a) shows the admission probability (i.e., the probability that user with latency-aware application could be admitted) as the function of the latency requirement L_{max} with local processing (LP) strategy and the proposed RI strategy. We notice from Figure 3.5(a) that the RI strategy significantly outperforms the LP strategy, since the insufficiency of computation resources can be made up for using communication resources in the RI strategy. The results indicate that more users with latency-aware applications can be supported using the concept of resource interchange. As well, we evaluate the minimal energy consumption E_{min} of UE under LP and RI schemes in Figure 3.5(b). Using LP strategy, E_{min} is always equal to the energy spent on total execution at UE. However, in RI, as the distance between UE and the cloud increases, E_{min} will increase at first and finally saturate at the value that equals the energy spent on total execution at UE. As expected, RI strategy outperforms LP. Particularly, when the distance is 20 m, E_{min} in the RI strategy is only 37.4% of that in LP. This result indicates that resource utilization can be significantly enhanced through the interchange of communication and computation resources.

3.1.2 Resource Mobility Enabled Interference Management Framework

In order to manage interference, we propose a framework based on resource mobility (i.e., resource mobility enabled architecture (RMEA)). In this architecture, all the resources in the networks including space resources, spectrum resources, storage resources, and computation resources are pooled together in the resource pool under a unified management. As well, the whole network is divided into a number of wireless resource aggregation regions as shown in Figure 3.2. A wireless resource aggregation region consists of a number of cells, where all the wireless resources can be aggregated and uniformly scheduled. In each region, there is a WRAC that implements the resource management. Simply put, when the data flows go through the network along a particular path, the resources will be aggregated and allocated by WRAC along this path from the resource pools. After data transmission is finished, all the resources will be sent back to the pools. Hence, by dynamically matching the varied traffic flows with guaranteed aggregated resources, on-demand services can be provided for the users. Figure 3.6 shows the functional entities of the RMEA (Figure 3.6(a)) and the corresponding functions (Figure 3.6(b)) that are executed within each functional entity. Figure 3.6(a) illustrates the four necessary functional entities

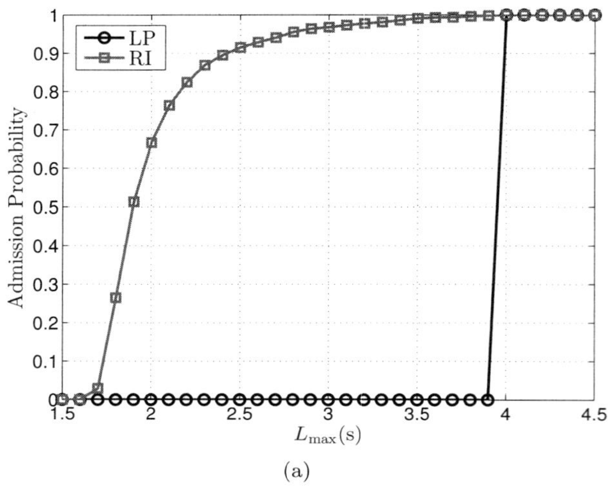

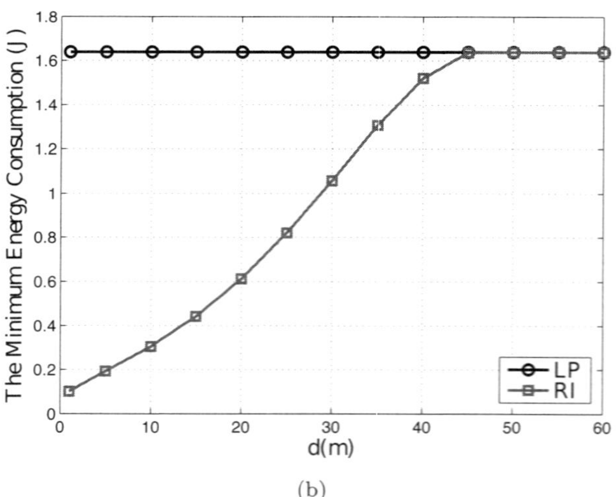

Figure 3.5: (a) Admission probability and (b) minimum energy consumption in a resource interchange scenario.

of RMEA: sensing, decision, reconfiguration and self-learning, which can be further divided into both terminal side and network side, respectively. The central controller WRAC manages all the available resources within the resource pool. Figure 3.6(b) depicts the four functions executed within the specific functional entities: sensing, interference management decision, self-learning, and interference management implementation. In the following, we will give a detailed illustration of each entity and the corresponding functions.

Sensing: This entity collects the information from both terminal side and network side and forwards it to the decision entity. In particular, the terminal-side-related information includes the QoS requirement and terminal-side-related resources and the interference status. The terminal-side-related resources include antenna resources, power resources, computation resources, and hardware resources. The interference status at the terminal side includes the interference strength and interference variation around the terminal. The network-side-related information is a collection of available resources and the network interference status, where the resources include space resources, spectrum resources and computation resources. The network interference status consists of the network interference distribution and the interference structure features.

Decision: This entity is a key module for realizing resource migration, aggregation, and interchange to enhance system capacity and improve QoS/QoE. It is used to make resource and interference management decisions based on the sensing results in both the terminal side and network side. The entity first collects all the resource and interference information and then executes the resource allocation and interference management operations with certain objectives. These operations consist of user scheduling and interference management, where users are scheduled with all the available resources within the the resource pool. For users allocated with the same resources, specific interference management techniques are selected to mitigate the created interference. Specifically, the decision entity includes an interference management technique database, which maintains some classical interference management techniques, such as interference migration, interference mitigation and interference avoidance.

Reconfiguration: This entity communicates with the decision entity periodically and receives the outcomes from the decision entity. The function of interference management implementation is executed within this entity. In particular, the function executes the corresponding resource allocation and interference management operations to guarantee that the resources can be dynamically utilized and the on-demand services can be provided.

Self-learning: This entity communicates with the other three entities, and performs the self-learning function to optimize the network performance. Specifically, it periodically receives resource and interference information coming from the sensing entity and predicts the resource and interference distribution over the whole network. After implementing the resource and interference management operation within the reconfiguration entity, the resulting network performance is input into the self-learning entity that evaluates the effect of the corresponding operations. The self-learning function further modifies the resource

Table 3.1: Differences Between RMEA and Several Typical Resource Management Methods.

Resource management methods	resource migration	resource aggregation	resource interchange	space resources	spectrum resources	computation resources	storage resources
RMEA	√	√	√	√	√	√	√
IEEE 1900.4 [7]	×	×	×	×	√	×	×
C-RAN [6]	√	√	×	×	×	√	×
H-CRAN [8]	√	√	×	√	×	√	×
CR networks [9]	×	×	×	×	√	×	×
Load balance [10]	√	√	×	×	√	×	×

and interference status, and serves as an input to the decision entity to adjust the interference management decision and update the interference management strategy database.

This architecture is backward compatible with several architectures (e.g., C-RAN [6] and IEEE 1900.4 [7]). Those functions can be executed in the corresponding entities. The differences between RMEA and several typical resource management methods in the three techniques and types of managed resources are presented in Table 3.1. From this table, we can see that RMEA is more pervasive to managing the resources in future wireless networks. Through designing RMEA, resource mobility becomes possible.

3.2 Interference Modeling for Heterogeneous Wireless Networks

Due to the scarcity of channel bandwidth, spectral resource reuse has been considered as a major way to improve network capacity. However, spectral resource reuse gives rise to cochannel interference, which may lead to performance degradation to the network. In a heterogeneous wireless network, multiple types of devices may coexist: high-power macrocell base stations (MBSs), different types of LPNs, and/or randomly deployed D2D transmitters. By treating each type of base stations (distinguished by transmit power and deployment intensity) as a tier, the cochannel interference in such a network can be classified into the following two types:

- Cross-tier interference: interference between different tiers of base stations that share the same spectrum resources,

- Cotier interference: interference among base stations in the same tier and/or interference between cellular transmissions and D2D transmissions.

Cochannel interference plays an important role in the achievable network performance, which should be carefully addressed. To understand the impact of

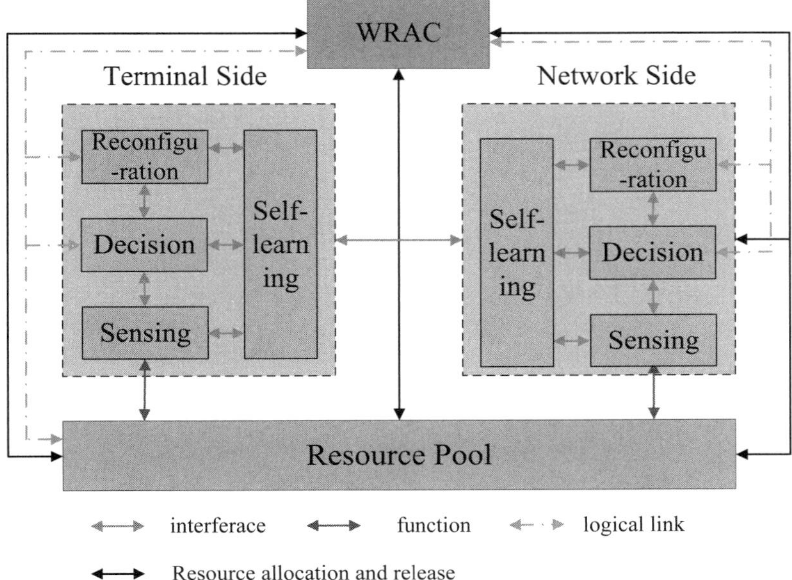

(a) Functional entities of RMEA

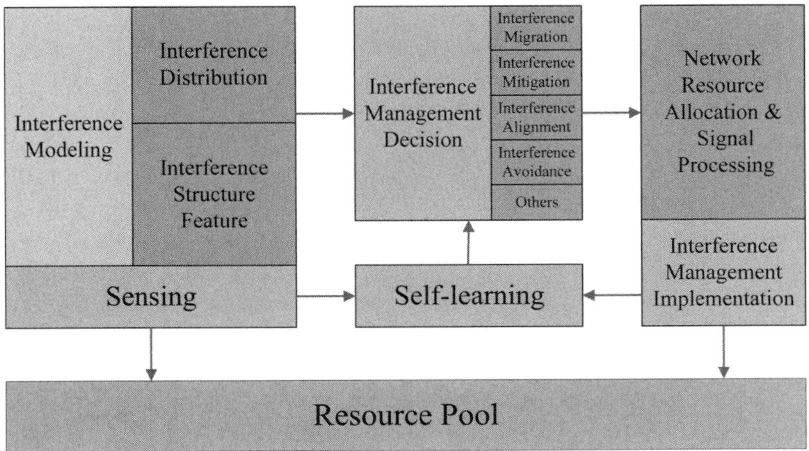

(b) Functions of each entity.

Figure 3.6: Illustration of RMEA.

interference on network performance, the first thing we should do is an accurate characterization of interference. Therefore, interference modeling is important in the design and deployment of networks, which is also necessary for the assessment of interference management techniques and protocols.

In this section, we introduce interference modeling methods that can be used for heterogeneous wireless networks, where we first give the channel model and then present two typical interference models: the physical interference model and protocol interference model. Then, a statistical interference model is proposed with the tool from stochastic geometry. The power of the model in characterizing interference caused by random deployment of interferers is illustrated with the interferers modeled by the PPP.

3.2.1　Channel Model

To characterize interference, the following three main propagation impacts should be taken into account [6]: deterministic path loss, large-scale fadingand small-scale fading. The deterministic path loss is used to model the attenuation of a signal traveling from a transmit antenna to a receive antenna. It is assumed that the deterministic path loss can be approximated by the following simple expression

$$P_r = \begin{cases} P_t K \left(\frac{d}{d_0}\right)^{-\alpha}, & d > d_0 \\ P_t K \left(\frac{d}{d_0} + 1\right)^{-\alpha}, & d \leq d_0 \end{cases}, \text{ or } P_r = P_t K \min\left(\left(\frac{d}{d_0}\right)^{-\alpha}, 1\right), \quad (3.3)$$

where P_t and P_r denote the transmit power and receive power, respectively, $K > 0$ is a constant value related to the average attenuation and antenna characteristics, d is the distance between the transmitter and receiver, d_0 is the reference distance for the antenna far-field, and $\alpha > 2$ is the path loss exponent depending on the propagation environment. For the near-field transmission when $d < d_0$, the path loss model is modified to avoid the unbounded behavior of power. A more sampled version of (3.3) is usually considered in most of the literature, where K and d_0 are set to be 1,

$$P_r = \begin{cases} P_t d^{-\alpha}, & d > 1 \\ P_t (d + 1)^{-\alpha}, & d < 1 \end{cases} \text{ or } P_r = P_t \min\left(d^{-\alpha}, 1\right), d > 0. \quad (3.4)$$

large-scale fading [7] is another fundamental element in wireless communications, which is caused by the shadowing of surrounding buildings and obstacles. It is usually modeled by a lognormal shadowing model, where the distribution of the signal amplitude X is given by

$$p_X(x) = \frac{10/\ln 10}{\sqrt{2\pi}\sigma_{x_{dB}} x} \exp\left[-\frac{(10\log_{10} x - \mu_{x_{dB}})^2}{2\sigma_{x_{dB}}^2}\right], x > 0, \quad (3.5)$$

where $\mu_{x_{dB}}$ is the mean value of $x_{dB} = 10 \log_{10} x$ in dB and $\sigma_{x_{dB}} \in [2,6]$ [dB] is the standard deviation of x_{dB}, also in dB. The mean power $\mu_{x_{dB}}$ is related to the path loss and the properties of buildings and obstacles. small-scale fadingrefers to the rapid envelop fluctuation in a distance or short period of time, which is caused by the constructive or destructive combination of multipath signals at the receive antenna. To describe the received signal magnitudes, some classical statistical distributions are commonly used: Rayleigh, Rician, and Nakagami-m distributions. Particularly, Rayleigh fading is most applicable to the case where no dominant component exists between the transmitter and the receiver. Rician fading is a more accurate model when there is one component that dominates (i.e., the line-of-sight component). Nakagami-m distribution can be used to model signal fading conditions ranging from severe to moderate, to weak fading through the parameter m. The probability density function (p.d.f.) of the signal amplitude X of all the three distributions are given by

$$f_X(x) = \begin{cases} \frac{x}{\sigma^2} \exp\left(-x^2/2\sigma^2\right), \, x \geq 0 & \text{Rayleigh \ fading} \\ \frac{x}{\sigma^2} \exp\left(\frac{-(x^2+v^2)}{2\sigma^2}\right) I_0\left(\frac{xv}{\sigma^2}\right), & \text{Rician \ fading} \\ \frac{2m^m x^{2m-1}}{\Gamma(m)\sigma^m} \exp\left(-\frac{m}{\sigma}x^2\right), & \text{Nakagami-}m \text{ \ fading} \end{cases} \tag{3.6}$$

where in Rayleigh fading, σ is the scale parameter, while in Rician fading, $I_0\left(z\right)$ is the modified Bessel function of the first kind with zero order, and $\sigma^2 = \frac{\Omega}{2(1+K)}$, $v^2 = \frac{K\Omega}{1+K}$ with K the ratio of power in direct path to the power in the scattered paths and Ω the total power from all the paths. The parameters m and σ in Nakagami-m fading is, respectively, given by $m = \frac{\mathbb{E}^2\left[X^2\right]}{Var[X^2]}$, and $\sigma = \mathbb{E}\left[X^2\right]$.

3.2.2   Typical Interference Models

In the current literature, two popular interference models are available: the *physical interference model* and *protocol interference model*, which are first formalized by Gupta and Kumar in their work on capacity of wireless ad hoc networks [8].

Let X_i denote the location of a transmitting node, and $X_{R(i)}$ the receiver of node X_i. The protocol interference model is based on a guard zone that defines a condition for successful communication over a given link. In particular, the transmission over link $X_i \rightarrow X_{R(i)}$ is successful as long as the following condition is satisfied

$$\mid X_j - X_{R(i)} \mid \geq (1 + \Delta) \mid X_i - X_{R(i)} \mid \tag{3.7}$$

where $\mid X_i - X_{R(i)} \mid$ and $\mid X_j - X_{R(i)} \mid$ represent the distances from the transmitting node X_j and X_i to the receiver node $X_{R(i)}$, respectively. The quantity $\Delta > 0$ is the guard zone radius specified by the protocol. It is worth noting that the protocol interference model only considers the deterministic path loss model and the same power level for all transmitters. For two active links $X_i \rightarrow X_{R(i)}$

and $X_j \to X_{R(j)}$ operating under the protocol interference model, we derive that both transmissions are successful only if

$$| X_{R(j)} - X_{R(i)} | \geq \frac{\Delta}{2} \left(| X_j - X_{R(j)} | + | X_i - X_{R(i)} |\right). \tag{3.8}$$

It shows that disjoint guard zones around receivers $X_{R(i)}$ and $X_{R(j)}$ of radius $\frac{\Delta}{2} | X_j - X_{R(j)} |$ and $\frac{\Delta}{2} | X_i - X_{R(i)} |$ are required to guarantee the successful transmissions of both links. Note that a small link length leads to a small guard zone, and then more links can be active simultaneously within the given area.

The protocol interference model ignores the interferers outside the guard zone of the receiver, which leads to an underestimation of the aggregate interference at a receiver. To improve this and relax the same power level constraint in the protocol interference model, the physical interference model was proposed. Let $\{X_k,\ k \in \mathcal{N}\}$ be the set of transmitting nodes operating on the same channel with power level $\{P_k\}$ for node X_k, $k \in \mathcal{N}$. The transmission from node X_i can be successfully received at node $X_{R(i)}$ only if the SINR at $X_{R(i)}$ is no less than the given SINR threshold γ; that is

$$\frac{\frac{P_i}{|X_i - X_{R(i)}|^\alpha}}{N_0 + \displaystyle\sum_{k \in \mathcal{N}, k \neq i} \frac{P_k}{| X_k - X_{R(i)} |^\alpha}} \geq \gamma, \tag{3.9}$$

where the simple deterministic path loss model given in (3.4) is employed, and N_0 is the thermal noise power, while the impact of large-scale fadingand small-scale fadingon the interference is ignored.

Note that knowledge of node locations is a precondition to employ the two typical interference models. However, node locations are not always available. In the heterogeneous wireless network, due to the random deployment of SAPs, D2D transmitters, and mobile users, it is impossible to know or predict the locations of all interfering nodes. This leads to great difficulty in accurately estimating the network interference. Thus, a statistical interference model is necessary to characterize the network interference distribution and evaluate the network performance.

3.2.3   Statistical Interference Model

The interference statistics in large-scale heterogeneous wireless networks depend on several sources of randomness, including node distribution, MAC protocol, channel fading, and the adopted different transmit power levels. First, interferers are scattered in the space and thus the interference statistics are related to the spatial distribution of concurrently transmitting nodes. Second, the set of transmitters operating on the same channel are affected by the MAC protocol that determines the active interferers at each moment. Third, channel fading distribution in combination with deterministic distance-dependent path loss plays an important role in signal attenuation. Finally, the different transmit power levels of interferers make interference more complicated. All the above factors determine the aggregate interference imposed at the given receiver.

3.2.3.1 Interference Characterization

Let $\Phi = \{X_i\} \subset \mathbb{R}^d$ denote the set of active transmitters of the heterogeneous wireless network, where $d = 2$ represents the Euclidian plane and $d = 3$ the Euclidian space. The aggregate interference at a receiver $X_{R(i)} \in \mathbb{R}^d$ can be expressed as

$$I\left(X_{R(i)}\right) = \sum_{X_k \in \Phi} P_k h_k l\left(\| X_k - X_{R(i)} \|\right), \qquad (3.10)$$

where P_k is the transmit power of node X_k, h_k is the channel fading power including the effect of shadowing and small-scale fading, and $l\left(\cdot\right)$ is the path loss function assumed to be dependent on the distance $\| X_k - X_{R(i)} \|$, with $\| \cdot \|$ representing the norm of $\mathbb{R}^d$. The SINR of the receiver $X_{R(i)}$ can be expressed as

$$\mathrm{SINR}\left(X_{R(i)}\right) = \frac{S\left(X_{R(i)}\right)}{N_0 + I\left(X_{R(i)}\right)}, \qquad (3.11)$$

where $S\left(X_{R(i)}\right) = P_i h_i l\left(\| X_i - X_{R(i)} \|\right)$ is the desired signal power received at node $X_{R(i)}$. From (3.10), we observe that the geometry of the locations of active transmitters plays a key role in determining the distribution of the interference and the SINR at each receiver. To characterize the effect of network geometry on the interference or SINR distribution, stochastic geometry has been proposed in the recent literature [9, 10]. As a powerful tool, stochastic geometry studies network interference by averaging over all potential geometrical patterns of nodes.

3.2.3.2 Poisson Point Process

With stochastic geometry, the uncertainty of nodes' locations can be represented by a certain spatial point process $\Phi = \{X_i\}$ on $\mathbb{R}^d$ where X_i denotes the node location and the node itself. We consider a simple case where no two nodes are at the same position. Then the point process can be written as a random set $\Phi = \{X_1, X_2, ..., X_N\}$ where the total node number N can be finite or infinite. At any moment t, a subset of nodes will be selected from set Φ according to the defined MAC protocol, which leads to a time-dependent active transmitter set $\Phi(t)$. In this case, the interference is also time-dependent. When the network interference is stationary, both in time and space, we can omit the time index and the node spatial location.

A widely used point process is PPP which is defined as follows. Let Λ be a locally finite measurement on certain metric space E, where Λ is defined as $\Lambda(A) = E\Phi(A)$ for Borel A, and $E \subseteq \mathbb{R}^d$. A point process Φ is Poisson on E if for any set A of E, the random variable (*r.v.*) $\Phi(A)$ (denoting the number of points in $\Phi(A)$) are Poisson. What's more, for all disjoint subsets A_1, ..., A_n of E, the $r.v.$'s $\Phi(A_i)$ are independent.

The merit of PPP model lies in its capability in deriving elaborate analytical expressions that provides a tractable way to quantify the effect of some important parameters on network performance. Now it is fairly well accepted for modeling both ad hoc networks [11, 12] and fixed infrastructures [13]. Although the PPP model is more reasonable for the unplanned small cell networks and randomly distributed ad hoc networks with uncoordinated MAC protocols (e.g., ALOHA). It is shown in [14] that even in one fairly regular macrocell deployment, the PPP model is about as accurate as the commonly used grid model.

3.2.3.3 Important Properties of PPP

1. *Campbell's theorem*: Define a measurable function $f(x) : \mathbb{R}^d \to [0, \infty]$, for a stationary PPP Φ with intensity λ, where the following equation is satisfied

$$\mathbb{E}\left[\sum_{x \in \Phi} f(x)\right] = \lambda \int_{\mathbb{R}^d} f(x)dx. \tag{3.12}$$

2. *Slivnyak's theorem*: Let δ_x be the Dirac measure on E, the law of PPP $\Phi - \delta_x$ on the condition that Φ has a point at x is the same as the law of Φ. That means the reduced Palm probability $P^{x!}$ of a PPP Φ is the distribution of Φ itself, written as $P^{x!} = P$.

With this property, it means that an additional point at the original o does not change the distribution of the other points of the PPP.

3.2.3.4 Application of PPP to Interference

Applying Slivnyak's theorem to (3.10), the total interference power received at any point $X_{R(i)}$ is equal to the aggregate interference at the origin, given by

$$I = I(o) = \sum_{X_k \in \Phi} P_k h_k l\left(\| X_k \|\right). \tag{3.13}$$

From Campbell's theorem, the mean value of the aggregate interference can be derived as

$$
\begin{aligned}
\mathbb{E}(I) &= \mathbb{E}\left[\sum_{X_k \in \Phi} P_k h_k l\left(\| X_k \|\right)\right] \\
&= P_k \lambda c_d d\mathbb{E}(h_k) \int_{\mathbb{R}^d} l\left(\| x \|\right) dx \\
&= P_k \lambda c_d d\mathbb{E}(h_k) \int_0^\infty l(r) r^{d-1} dr, \tag{3.14}
\end{aligned}
$$

where $c_d =\mid b(o, 1) \mid$ is the volume of the unit ball in the d-dimensional space. Specifically, $c_2 = \pi 1^2 = \pi$, $c_3 = \frac{4}{3}\pi 1^3 = \frac{4}{3}\pi$. For $\mathbb{E}(h_k) = \mu$ and $l(r) = (r^{-\alpha}, 1)$, we have

$$\begin{aligned}\mathbb{E}\left(I\right) &= P_k \lambda c_d d\mu \int_0^\infty \min\left(r^{-\alpha}, 1\right) r^{d-1} dr \\ &= P_k \lambda c_d d\mu \left(\frac{1}{d} + \frac{r^{d-\alpha}}{d-\alpha} \Big|_1^\infty\right).\end{aligned} \tag{3.15}$$

Note that when $\alpha \leq d$, the integration in (3.15) diverges to infinity, while when $\alpha > d$, the mean interference is given by

$$\mathbb{E}\left(I\right) = P_k \lambda c_d \mu \left(1 + \frac{1}{\alpha - d}\right). \tag{3.16}$$

The PPP model provides a feasible way to characterize the network interference distribution and the corresponding SINR distribution. It is most useful in large-scale heterogeneous wireless networks, where multiple types of interferers contribute to the aggregate interference. By modeling the positions of each type of base stations or D2D transmitters as a distinct PPP, the aggregate interference can be expressed and characterized by employing the properties of the PPP model.

3.3 Techniques for Interference Management

In this section, we introduce some interference management techniques (e.g., interference migration, interference mitigation, and interference alignment) under the interference management framework described in Section 3.1. In particular, interference migration corresponds to resource migration and aggregation, where the interference can be migrated by adjusting the transmit power of two aggregated networks. interference mitigation is related to resource migration, where the interference can be reduced or avoided by assigning different resources to different interferers. Interference alignment corresponds to resource interchange, where the interference is aligned at the expense of computation cost at the transceivers (i.e., transmit beamforming and receiver shaping).

3.3.1 Interference Migration

In heterogeneous networks, both cotier and cross-tier interference exist, which may deteriorate the system performance. Moreover, due to the random deployment of small cells (e.g., femtocells) and unbalanced load distribution, interference in different regions presents significant dissimilarity (i.e., some regions have severe interference but interference of the other regions is slight). Therefore, from a whole network point of view, the distribution of interference is nonuniform. By leveraging dual connectivity, a UE can simultaneously connect with two different base stations to exploit the nonuniform interference distribution. In particular, the user can shift transmission power among different connections

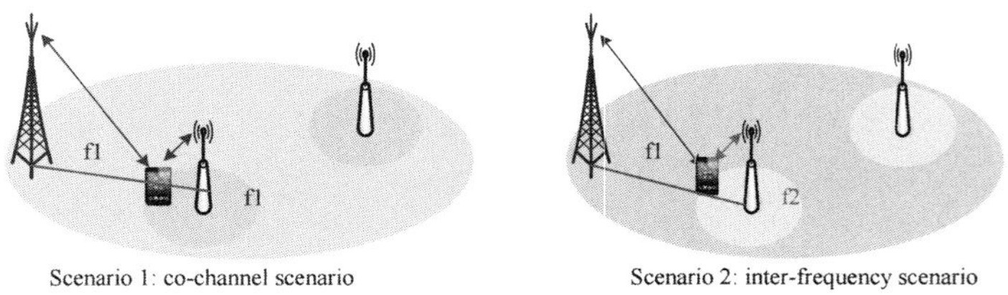

Figure 3.7: Cochannel and interfrequency deployment scenarios.

by adjusting the transmission rate of each connection, thereby transferring interference from the severe region to the slight region.

There are two deployment scenarios, namely, cochannel scenario and interfrequency scenario, as shown in Figure 3.7. In the first scenario, the UE is connected to macro- and small cells on the same carrier frequency, while in the second scenario, the UE is connected to macro- and small cells on different carrier frequencies. In both scenarios, the macro and small cells are connected via nonideal backhaul.

We use a downlink data flow on a 4G network as an example to explain how data is sent via macro- and small cell base stations to the UE. User plane data from the core network (CN) is first transferred to the macro eNB, which operates as the master eNB (MeNB). In the macro eNB, the data flow is split into two parts. Some data is transmitted via the macro (PCell) to the UE, while other data is transferred over the X2 interface to the small cell eNB, which operates as the secondary eNB (SeNB), and transmitted to the UE via the corresponding cell (SCell).

Although in theory the roles of master and secondary eNB do not depend on the eNB's power class and can vary among UEs, we assume that the MeNB is always a macro eNB while the SeNB is always a small cell eNB. The X2 interface imposes latencies of a few milliseconds to several tens of milliseconds depending on the implementation. In alignment with 3GPP assumptions, the MeNB and SeNB are assumed to have independent medium access control (MAC) entities and physical layer processing. This implies that the macro- and the small cell eNBs each decide how to schedule data for the UE. Similarly, an independent hybrid automatic repeat request (HARQ) and link adaptation are used for the PCell and SCell transmissions in line with basic carrier aggregation (CA) assumptions. When the UE has multicarrier transmission capability for the uplink, it can feedback separate channel state information (CSI) and HARQ acknowledgments (ACK) to the macro- and small cell eNBs. Once the data packets have been decoded successfully by the UE, they are reordered and delivered to higher layers. It is therefore obvious that the performance depends on multiple factors, where the design of RRM algorithms for deciding serving cell(s) for the UEs, packet scheduling, power control, and flow control between the evolved eNBs over the X2 interface are of particular importance.

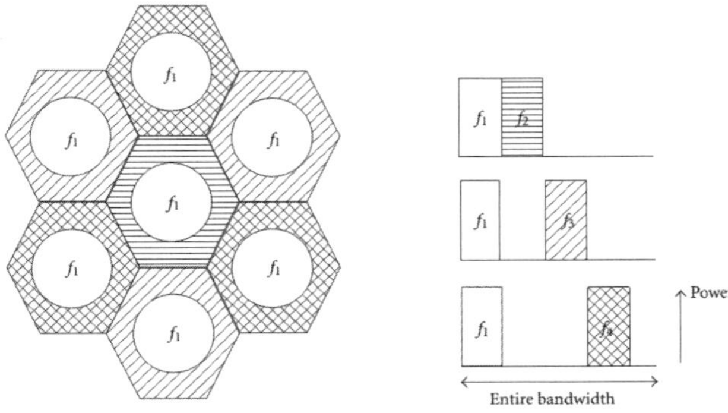

Figure 3.8: Illustration of fractional frequency reuse.

3.3.2   Interference Mitigation

Typical interference mitigate schemes can be roughly categorized into three types: frequency assignment, power control, and antenna schemes. Joint schemes are also possible by combining two or more schemes. We will next briefly overview these schemes.

Spectrum Assignment

Intercell interference (ICI) is a major problem in heterogeneous networks. As the cell-edge performance is particularly susceptible to ICI, improving its performance is an important aspect of systems design. To mitigate the interference in heterogeneous networks, a BS assigns its UEs a spectrum with limited or no interference with a neighboring BS. The choice between a dedicated or cochannel deployment is implemented with considerations such as the amount of spectrum available and density of BSs in a specified region. Specifically, there are two frequency reuse methods: fractional frequency reuse (FFR) and soft frequency reuse (SFR). In FFR, the total available bandwidth is split into multiple parts, for example four parts $\{f_1, f_2, f_3, f_4\}$ as shown in Figure 3.8. One part (e.g., f_1) is allocated to the cell center of each cell, while the other parts (e.g., f_2, f_3, f_4) are allocated to the cell edge of different cells. An interference avoidance factor (IAF) based on fairness can be introduced to balance the number of UE and resource block (RB) allocations. That is, IAF is used to determine the number of RBs which should be avoided in neighboring cells.

Although a lower ICI is achieved in FFR by allocating different frequency segments to adjacent cell-edge regions, this strict no-sharing policy may under-utilize available frequency resources in certain situations. In order to avoid the high ICI levels associated with universal frequency reuse, while providing more flexibility to the FFR scheme, an SFR scheme has been proposed in which the entire bandwidth can be utilized. As in FFR, the available bandwidth is divided into orthogonal segments and each neighboring cell is assigned a cell-edge band.

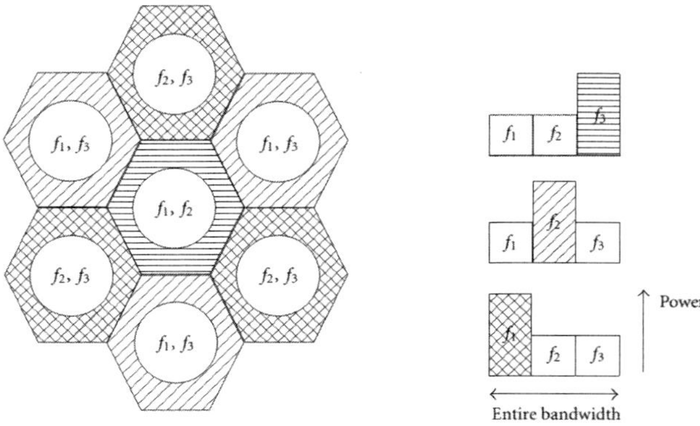

Figure 3.9: Illustration of soft frequency reuse.

In contrast to FFR, a higher power is allowed on the selected cell-edge band, while the cell-center UE can still have access to the cell-edge bands selected by the neighboring cells, but at a reduced power level. In this way, each cell can utilize the entire bandwidth while reducing interference to the neighbors. An example SFR scheme is illustrated in Figure 3.9. Note that this scheme can improve the SINR of the cell-edge UEs, while degrading the SINR of the cell-center UE. This degradation is due to the overlap in frequency resources between the cell-edge band of the neighboring cells and the cell-center band of the serving cell. However, as mentioned earlier, the cell-edge performance improvement is almost linear while the degradation to the cell-center UE is logarithmic. In SFR, the power ratio between the cell-edge band and the cell-center band can be an operator-defined parameter, thereby increasing the flexibility in system tuning.

Downlink Power Control

The radiated power transmitted by BSs comprises pilot power and traffic power (consisting of signaling and data). The effect of interference on other BSs is dependent on these two power levels. A high pilot power will result in a large cell coverage area, which consequently has higher chances of causing interference. There is a need to optimize the transmit power in BSs to avoid interference while maintaining a certain level of QoS. Figure 3.10 illustrates a scenario where BS1 and BS2 are deployed in two flats, Flat A and Flat B, serving SUE1 and SUE2, respectively, with their pilot power levels overlapping each other (solid circles). In this scenario, BS1 employs a power control mechanism that reduces its pilot power (i.e., coverage area (dotted region)) thus preventing cotier interference to SUE2 and BS2 and cross-tier interference with the MUE. Power control, however, is not only restricted to the BSs as UE can also optimize its power levels or assist its BSs to reduce interference to neighboring BSs and other UE.

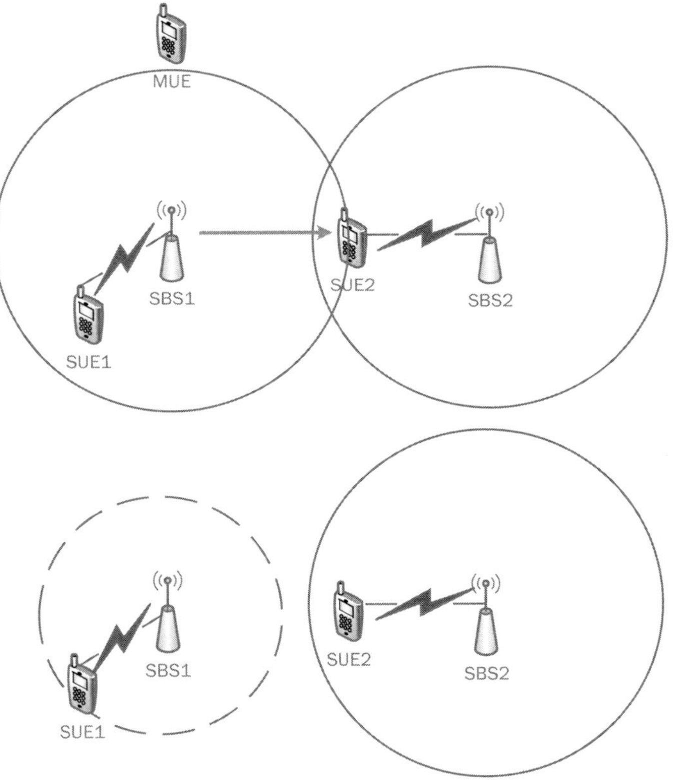

Figure 3.10: Illustration of power control.

Uplink Power Control

Uplink power control is a key radio resource management feature in 4G cellular communication systems, which consists of an open-loop scheme handled by the user equipment and closed-loop power corrections determined and signaled by the network. Uplink power control is usually used to provide an adequate transmit power to the desired signals to achieve the necessary quality, minimize the interference to other users in the system, and maximize the battery life of the mobile terminal. In order to achieve these goals, uplink power control has to adapt to the radio propagation channel conditions, including path loss, shadowing, and fast fading fluctuations, while limiting the interference effects from other users, within the cell and from neighboring cells. With open-loop power control, the transmit power is given by $P = \min\{P_{\max}, 10 \times \log_{10} M + P_0 + \alpha \times P_L\}$, where $P_{\max}$ is the maximum transmit power, M is the e number of physical resource blocks, P_0 is cell/UE specific parameter, which is used to control SNR target and is signaled by the radio resource control (RRC), α is the path loss compensation factor, and P_L is the downlink path loss estimate. The LTE closed-loop power control mechanism operates around an open-loop point of operation. The UE adjusts its uplink transmission power based on the TPC commands it receives from the eNB when the uplink power setting is performed at the UE using open-loop power control.

Antenna Schemes

Beam directivity of the antennas, both in BS and UE, can be exploited to avoid interference in heterogeneous networks. Typical antenna schemes are implemented that allow the BS to direct their beams to specific UE while creating a null in other zones, thus canceling the interference to a greater extent, as shown in Figure 3.11. In 4G LTE, an important application of beamforming is coordinated multipoint (CoMP). BSs deployed by the same operator are interconnected through an interface, such as the X2 interface in LTE-A provided for the information interchange. With efficient coordination, the signal received from other cells can be used constructively and it can be an information resource rather than an interference source. This is the basic idea behind CoMP. The 3GPP LTE-A standard highlights two approaches for CoMP that mainly differ on the required network signaling overhead. The former LTE CoMP scheme is called coordinated scheduling/coordinated beamforming (CS/CB). It is mostly based on the sharing of information provided by each terminal to its BSs, as well as other information exchanged among BSs, to contribute to a joint decision. In this context, the data addressed to a specific UE is transmitted by a single BS, while other BSs can use the same radio resources (i.e., the same PRBs) to transmit to angularly separated UE using beamforming. The latter CoMP scheme proposed by the 3GPP is called joint processing/joint transmission (JP/JT). When a higher signaling overhead is supported, more sophisticated cooperation can be applied. In JP/JT, data transmission for each UE is done simultaneously by multiple TPs. Hence, the UE can combine multiple copies of the signal

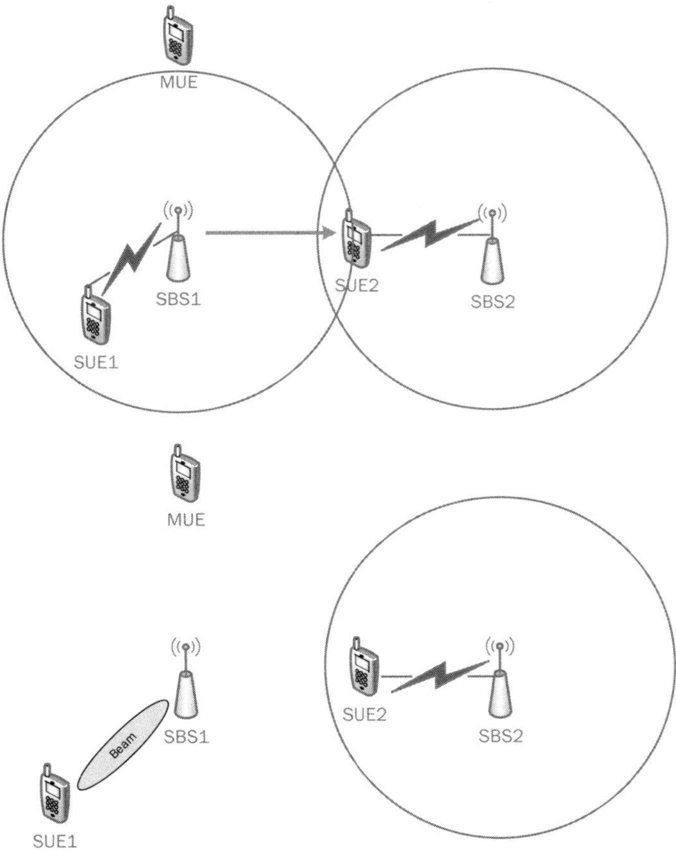

Figure 3.11: Illustration of antenna selection.

to improve the received signal quality. The waste of resources due to multiple transmissions is compensated for by beamforming that allows to spatially reuse the same resources to transmit toward angularly separated UEs.

To sum up, the problems associated with mitigating interference in heterogeneous networks are largely dependent on its tiered architecture and random deployment where there is no central coordination between its neighboring small cells and the macrocell network. Importantly, information about the radio environment, such as characteristics of all the interfering signals, if known, can be controlled to help mitigate interference in small cells. An ideal small cell interference mitigation technique is the one that is aware of the interfering signals and takes into consideration the best deployment criteria to suit subscribers' needs while efficiently utilizing the network operators' resources.

3.3.3 Interference Alignment

An interference channel is where several sender-receiver pairs share a common media so that transmission of information from one sender to its corresponding receiver interferes with communications between the other pairs. Providing each user with exclusive access to a fraction of wireless resources is a traditional method to handle the interference, for example, time-division multiple access (TDMA) and frequency-division multiple access (FDMA). In TDMA, users take turns one at a time to use the frequency band for transmission periodically, while in FDMA, each user is allocated a small portion of frequency band without overlap. Although there is no interference in these orthogonal access schemes, the performance is far from the capacity of interference channel.

Other interference management approaches include treating the weak interference as noise and decoding the strong interference. When interference is weak, the receiver can treat it as noise because it is known that introducing structure in weak interference does not help. However, heterogeneous wireless networks are often interference-limited. When the interference is much stronger than the signal, the receiver can decode the interference first and then subtract it from the received signal to decode the desired signal. However, it is less common in practice due to complexity of multiuser detection.

Interference alignment (IA) is an emerging technique in interference management, which is a linear precoding technique that attempts to align interfering signals in space, time, or frequency. IA can achieve optimal degrees of freedom (DoFs), which can be interpreted as the number of resolvable signal space dimensions. In other words, it can reach the capacity of the interference channel at very high signal-to-noise ratio (SNR).

Space Dimension

As MIMO is becoming one of the most prominent techniques in 3G, 4G, and future 5G systems, most research work on IA address space dimension. IA in space dimension can be achieved by precoding and decoding across multiple antennas. In a K-user interference network, each transmitter only sends data streams to its corresponding receiver, and the transmission of the users interferes with each other, which will affect the sum rate significantly if no interference management techniques are applied. Fortunately, IA can eliminate interference among users completely. By cooperatively designing the precoding matrices, interference can be constrained into certain subspaces at the unintended receivers. As such, interference can be eliminated at each receiver through the decoding matrix, which is orthogonal to the direction of interference, and the desired signal can be recovered in the other half of the signal space free of interference. Thus, "half the cake" in a manner of speaking can be obtained by each user.

We first consider a three-user classical IA structure as shown in Figure 3.12 to clarify the basic idea. In the IA interference channel, there are three users, each of which is equipped with two antennas. One data stream is sent by each transmitter to its corresponding receiver. Taking the first user, for example, its

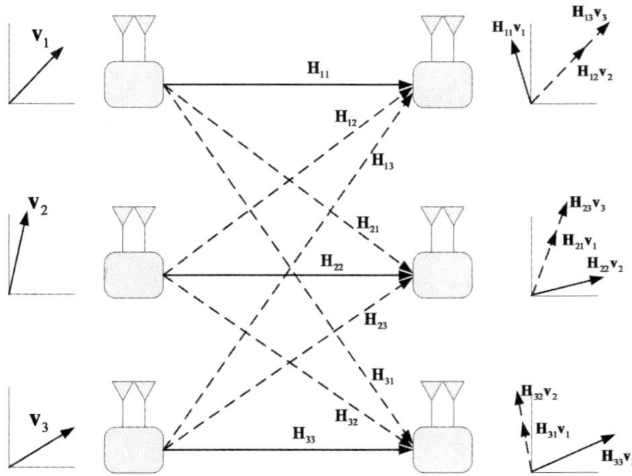

Figure 3.12: Interference alignment on the three-user MIMO interference channel with two antennas at each user.

received signal at the two antennas can be expressed as

$$\mathbf{y}_1 = \mathbf{H}_{11}\mathbf{v}_1 x_1 + \mathbf{H}_{12}\mathbf{v}_2 x_2 + \mathbf{H}_{13}\mathbf{v}_3 x_3 + \mathbf{n}_1,$$

where $\mathbf{H}_{kj}$ is the channel matrix from the jth transmitter to the kth receiver, $\mathbf{v}_k$ is the unitary precoding vector of the kth transmitter, x_k is the transmitted data stream of the kth user, and n_k is the additive white Gaussian noise (AWGN) vector at the kth receiver.

Because interference at the first receivers are aligned by using IA, we can assume that $\mathbf{H}_{11}\mathbf{v}_1 = [a\ b]^T$, $\mathbf{H}_{12}\mathbf{v}_2 = \beta[c\ d]^T$, and $\mathbf{H}_{13}\mathbf{v}_3 = \gamma[c\ d]^T$. When the AWGN n_1 is ignored, the received signal at the two antennas of the first receiver can be expressed as

$$\mathbf{y}_1[1] = ax_1 + c(\beta x_2 + \gamma x_3), \tag{3.17}$$

$$\mathbf{y}_1[2] = bx_1 + d(\beta x_2 + \gamma x_3). \tag{3.18}$$

In the above linear system, there are two linear equations (3.17) and (3.18) with three variables, x_1, x_2, and x_3. When IA is not applied, it cannot be solved because there are more variables than equations. If IA is performed, the undesired variables x_2 and x_3 are aligned, and we can denoted $x_4 = \beta x_2 + \gamma x_3$. The linear system of (3.17) and (3.18) can be rewritten as

$$\mathbf{y}_1[1] = ax_1 + cx_4, \tag{3.19}$$

$$\mathbf{y}_1[2] = bx_1 + dx_4. \tag{3.20}$$

There are two variables, x_1 and x_4, in these two equations, and thus the transmitted stream x_1 can be recovered by the first receiver.

For the general K-user MIMO interference channel, interference can be perfectly eliminated when the following conditions are satisfied

$$\mathbf{U}_k^\dagger \mathbf{H}_{ki} \mathbf{V}_i = \mathbf{0}_{d_k \times d_i}, \quad \forall i \neq k, \tag{3.21}$$

$$\mathrm{rank}\left(\mathbf{U}_k^\dagger \mathbf{H}_{kk} \mathbf{V}_k\right) = d_k, \quad k \in \{1, 2, \ldots, K\}, \tag{3.22}$$

where d_k and d_i are the number of data streams of user k and user i, respectively. The first condition guarantees interference at all the receivers can be completely eliminated, while the second condition ensures that the desired d_k data streams of the kth user should not be zero-forced at its receiver. The second condition can be satisfied almost always when the channel matrices do not have any special structures, and thus only the first condition should be focused on to realize IA.

When no symbol extension is utilized, the total DoFs are upper bounded by $M + N - 1$, where M and N are the number of antennas at each transmitter and receiver, respectively. However, when symbol extensions are exploited by using the coding over parallel time or frequency slots, the total achievable DoFs can grow linearly with the number of users K, and every user can achieve approximately one half of the capacity that is achieved in the one point-to-point channel without interference.

Time Dimension

In time dimension, IA is achieved through either propagation delays or coding across channels of time-varying fading, and a typical example is shown in Figure 3.13, in which the propagation delay of each link is adjusted to align interferences in half of the time slots. Time IA is usually leveraged in the scenarios where the number of antennas at each transceiver is not enough (e.g., single-input and single-output (SISO) based IA systems). A special kind of time IA is ergodic IA, in which the coefficients from the interfering users can be canceled in two well-chosen time slots, due to the complementary nature of these two channel states. Although its coding is very simple, long delay waiting for the desired channels is needed in ergodic IA, which makes it somewhat impractical.

Frequency Dimension

To achieve IA in the frequency dimension, the doppler shifts or coding across multiple carriers with frequency selective fading can be exploited. When there are no adequate antennas in the system, the frequency IA can be utilized. Besides, in long term evolution advanced (LTE-A) systems, orthogonal frequency-division multiple access (OFDMA) is adopted as the key technique of multiple access, by assigning subsets of sub-carriers to individual users. Due to the multifrequency nature of OFDMA, frequency IA is suitable to be utilized in orthogonal frequency-division multiplexing (OFDM) and OFDMA systems with a single antenna at each node. As the basic principles of time IA and frequency IA are almost the same, there no fundamental distinction exists between the IA techniques in time and frequency dimensions.

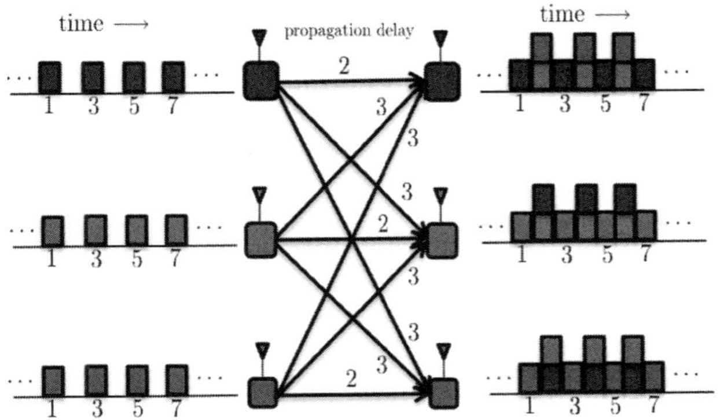

Figure 3.13: IA in time dimension on the three-user interference channel.

3.4 Technical Challenges in Interference Management

In the heterogeneous wireless networks, there still are some technical challenges to be addressed. For interference modeling, although the statistical model is powerful in interference characterization, it still has the following shortcomings. First, it can only reflect the average behavior of the network interference where all the network entities should have the same priority. Second, it can only model the network with general interference techniques while interference under carefully designed interference strategies cannot be exactly characterized. Third, the statistical interference model cannot reflect interference variation with time, and thus, it cannot characterize the time dynamics of interference. Therefore, other interference modeling methods should be exploited where the trade-off between computational complexity and accuracy should be taken into account.

For interference management techniques, some issues should be solved to apply IA to more practical scenarios. First, as we analyzed above, IA can solve the interference management problem with less antennas. Thus it is natural and important for us to know at least how many antennas are needed to achieve perfect IA. This is actually the problem of the feasibility condition. Second, the global CSI should be available at each node to calculate the solutions of IA. This is difficult to achieve, and the overhead of CSI feedback is extremely high. In addition, the closed-form solution of IA is usually difficult to obtain, especially when there are a large number of users in the network. Thus to obtain the IA solutions with low computational complexity is vital when IA is utilized. Furthermore, IA will fall short of the theoretical maximum when SNR becomes lower, and the resources in the wireless system should be fully exploited to improve the sum rate of IA networks especially at low SNR.

REFERENCES

[1] J. Andrews, S. Buzzi, W. Choi, S. Hanly, A. Lozano, A. Soong, and J. Zhang, "What will 5G be?" *IEEE J. Sel. Areas Commun.*, vol. 32, no. 6, pp. 1065–1082, Jun. 2014.

[2] J. Wen, M. Sheng, and Y. Zhang, "Dynamic hierarchy resource management for heterogeneous cognitive network," *Journal on Communications*, vol. 33(1), pp. 107–113, 2012.

[3] J. Wen, M. Sheng, Y. Zhang, and X. Wang, "On the performance of heterogeneous wireless networks with heavy tail traffic," in *Proc. IEEE ICCC*, Aug. 2013, pp. 286–291.

[4] X. Ma, M. Sheng, and Y. Zhang, "Green communications with network cooperation: A concurrent transmission approach," *IEEE Commun. Lett.*, vol. 16, no. 12, pp. 1952–1955, Dec. 2012.

[5] Y. Wang, M. Sheng, X. Wang, L. Wang, W. Han, Y. Zhang, and Y. Shi, "Energy-optimal partial computation offloading using dynamic voltage scaling," in *Proc. IEEE ICC Workshop on Cloud-Processing in Heterogeneous Mobile Communication Networks (IWCPM)*, Jun. 2015, pp. 2695–2700.

[6] M. D. Yacoub, *Foundations of Mobile Radio Engineering.* CRC Press, 1993.

[7] T. S. Rappaport, *Wireless Communications.* Prentice Hall PTR, 1996.

[8] P. Gupta and P. R. Kumar, "The capacity of wireless networks," *IEEE Trans. Inf. Theory*, vol. 46, no. 2, 2000.

[9] M. Haenggi, J. G. Andrews, F. Baccelli, O. Dousse, and M. Franceschetti, "Stochastic geometry and random graphs for the analysis and design of wireless networks," *IEEE J. Sel. Areas Commun.*, vol. 27, no. 7, pp. 1029–1046, Sept. 2009.

[10] D. Stoyan, W. Kendall, and M. J., *Stochastic Geometry and Its Applications.* John Wiley and Sons, 1996.

[11] S. Weber, J. G. Andrews, and N. Jindal, "The effect of fading, channel inversion, and threshold scheduling on ad hoc networks," *IEEE Trans. Inf. Theory*, vol. 53, no. 11, pp. 4127–4149, Nov. 2007.

[12] S. Weber, J. G. Andrews, and N. Jindal, "An overview of the transmission capacity of wireless networks," *IEEE Trans. Commun.*, vol. 58, no. 12, pp. 3593–3604, Dec. 2010.

[13] H.-S. Jo, Y. J. Sang, P. Xia, and J. G. Andrews, "Heterogeneous cellular networks with flexible cell association: a comprehensive downlink sinr analysis," *IEEE Trans. Wireless Commun.*, vol. 11, no. 10, pp. 3484–3495, Oct. 2012.

[14] J. G. Andrews, F. Baccelli, and R. Ganti, "A tractable approach to coverage and rate in cellular networks," *IEEE Trans. Commun.*, vol. 59, no. 11, pp. 3122–3134, Nov. 2011.

Chapter 4

Interference Migration: Leveraging Resource Heterogeneity

4.1 Introduction

To improve indoor coverage and enhance network capacity, femtocells are widely used due to their support for plug-and-play deployment that allows users to deploy them anytime when needed [1]. Although there are advantages to using femtocells, mass deployment of femtocells also incurs problems. On the one hand, the strong cotier and cross-tier interference that is introduced may deteriorate system performance.n the other hand, due to the random deployment of femtocells and the unbalanced load distribution, interference in different regions presents significant dissimilarity (i.e., some regions have severe interference but interference in the other regions is slight). In other words, the distribution of interference is nonuniform. To satisfy the QoS, users in the severe interference region usually have to increase the transmission power at the cost of decreasing energy efficiency (EE).

There have been a number of prior works addressing different aspects of interference management in HetNets scenario [2]. The main interference management techniques in the current literature can be summarized into two categories: interference cancellation and interference mitigation. With interference cancellation, interference from the received signal is canceled by decoding the desired information [3]. According to the method of cancellation, interference cancellation can be further divided into the following two kinds: successive interference cancellation (SIC) [4] and parallel interference cancellation (PIC) [5]. With SIC, one user is detected per stage while with PIC, all users can be detected simultaneously within one stage. Both SIC and PIC require a good understanding of interference signals from all interferers. This means that interference cancellation techniques can only deal with interference that can be received directly and can be centrally controlled.

In contrast with interference cancellation, the aim of interference mitigation is to avoid interference from the source [1] rather than cancel it in the receiver. The main approaches to mitigate interference are mainly focused on power control, fractional frequency reuse (FFR), spectrum splitting, and so forth [6]. As a key method for interference mitigation, power control has been used in [7] where a distributed utility-based approach is proposed to decrease the cross-tier interference subject to the SINR values. Spectrum splitting is seen as an effective method in managing interference, and 3GPP suggested that different portions of spectrum can be allocated for different femtocells [8]. In [9], an interference avoidance scheme is proposed to mitigate the inter-cell interference at the cell edges.

Apart from traditional network performance metrics, such as success probability and network throughput, EE is now attracting more and more attention due to the increasing awareness of environmental protection and energy consumption [10]. Currently, most of the research focuses on EE in radio access technologies, which contributes about 60% to 80% of the whole network energy consumption [11]. More important, the whole energy consumption of a user equipment relies on the limited battery energy that determines its lifetime. This implies that EE is crucial to the usability of user equipments [12]. Furthermore, more than 60% of users complain that the limited battery capacity is the most important factor impeding their use of data-hungry apps. Therefore, how to improve the EE is becoming increasingly important for green communication. Interestingly, if we can effectively utilize this nonuniform interference distribution property in interference-limited wireless networks, EE performance can be improved.

Existing studies on interference cancellation or mitigation have only looked at interference of nearby cells or users, and little attention has been paid to dealing with the nonuniform distribution of interference among different spatially separated regions. Fortunately, the advent of the mobile hotspot (MH), which has been widely investigated from the perspective of load balance and EE [13–15], brings us a new opportunity to address the nonuniform interference distribution problem via concurrent transmission. In particular, concurrent transmission allows the user to connect to different base stations and transmit the traffic via these connections simultaneously. By adjusting the transmission rate of each connection, the user can shift transmission power among different connections.

Inspired by nonuniform interference distribution and concurrent transmission, in this chapter, we propose a novel strategy called *interference migration* to enhance user EE by transferring interference from the severe region to the slight region. It is worth noting that the proposed interference migration strategy can be jointly combined with interference cancellation or interference mitigation strategies, so as to further improve network performance. Specifically, one can first utilize interference cancellation or interference mitigation strategy to cancel or reduce strong interference in the interference-severe region. For interference that cannot be canceled, one can further utilize our proposed interference migration strategy to transfer a partial amount of interference to the interference slight region. In this chapter, we will explore the effect of interference migration on the performance of the universal-frequency-reuse HetNet. Note that the concurrent transmission can be divided into two different types: within single RAT and among multiple RATs. We first consider interference migration via concurrent transmission within the single RAT and then show how to extend the concurrent transmission among multiple RATs. The concurrent transmission strategy is related to the existing practical techniques, such as carrier aggregation (CA) and LTE-WiFi Aggregation (LWA) strategies. For instance, CA has been introduced in LTE-A Release 10 and beyond, where multiple LTE component carriers are utilized to support the high data rates. LWA is a multiple RATs CA, where the data traffic is divided into different subflows and transmitted via different LTE interface and WiFi interface. As such, the licensed spectrum and unlicensed spectrum are aggregated to satisfy the traffic demands. Both strategies utilize the concept of concurrent transmissions with CA and LWA with the single RAT scenario and multiple RATs scenario, respectively.

4.2   HetNet Model with Single RAT

We consider the uplink transmission in a three-tier HetNet, which consists of a macrocell base station, K femtocell base stations, and an MH, as shown in Figure 4.1.[1] The universal-frequency-reuse scheme is adopted where the macrocell base station and its overlaid K femtocell base stations operate in the same spectrum. The spectrum is divided into different component carriers, each of which is the minimum resource granularity for cellular networks. The MH is a novel multiprotocol device deployed in HetNet [16–19]. Specifically, MH creates a WiFi hotspot and connects a WiFi user to the Internet via cellular backhaul. Therefore, the traffic of such a user will be transmitted via a two-hop link. The first hop is from the user to MH via WiFi connection, while the second one is from MH to base station via cellular interface.

To perform the concurrent transmission in the three-tier HetNet, we introduce a novel user equipment called concurrent transmission user equipment (CUE). In contrast to the traditional macrocell user equipment and femtocell user equipment, the CUE can establish the connections via base station and MH simultaneously. Consequently, the traffic of CUE can be split into subflows and transmitted via different interface concurrently. As shown in Figure 4.1, the CUE and MH access to different base stations. When concurrent transmission is executed, the CUE will distribute data packets to different interfaces with certain splitting probabilities that determine the transmission rate of each connection. Then, the traffic will be transmitted via different base stations, carried

---

[1]It is worth noting that the proposed interference migration strategy can be applied to both uplink and downlink directions, and we use the uplink direction as an example.

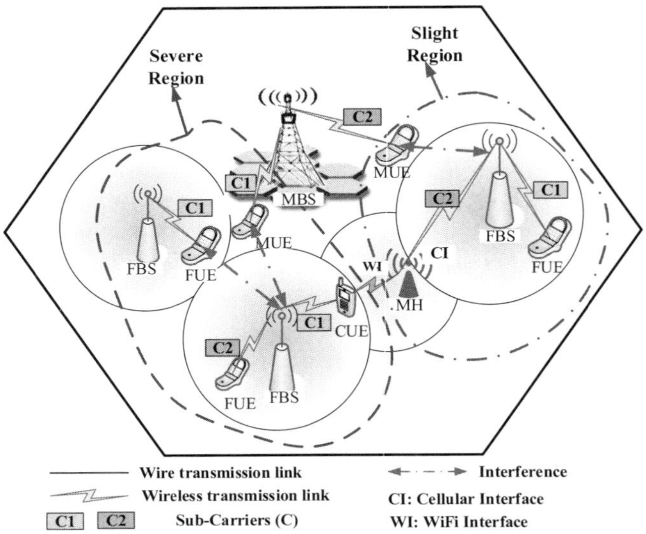

Figure 4.1: A three-tier HetNet scenario with single RAT.

through various networks, and finally merged in the destination.

4.2.1 Interference Distribution Model

To characterize the interference distribution, we first assume the Rayleigh fading channel model. We define $S(i)$ as the serving base station of user i, and $G_{i,S(j)}$ as the channel power gain from user i to the serving base station of user j. We refer to user j as an interference neighborhood of user i if and only if $G_{j,S(i)}P_j > \hat{\gamma}$, where P_j is the transmit power of user j, and $\hat{\gamma}$ is the predefined interference threshold. Furthermore, we define the set of all interference neighborhoods of user i as $\mathbb{N}_i$. To simplify the expression, we normalize the channel power gain and noise power (i.e., $G'_{j,S(i)} = G_{j,S(i)}/G_{i,S(i)}$ and $n'_i = n_i/G_{i,S(i)}$), where n_i is the background noise power. Accordingly, the normalized interference plus noise of user i is defined as

$$v_i = \sum_{j \in \mathbb{N}_i} G'_{j,S(i)}P_j + n'_i. \tag{4.1}$$

Let $\mathbb{N}_C$ and $\mathbb{N}_M$ denote the interference neighborhood sets of CUE and MH. Because CUE and MH can access to different base stations that are deployed in different locations, $\mathbb{N}_C$ and $\mathbb{N}_M$ may distribute in different spatial regions. Furthermore, due to the random deployment of femtocell base stations and unbalanced load distribution, the interference may exhibit a large difference between these two sets. Therefore, the interference of different spatial regions may distribute nonuniformly. To depict the nonuniform interference distribution, we define a term called interference intensity index $\zeta_{C,M}$ as

$$\zeta_{C,M} = \frac{\nu_C}{\nu_M} = \frac{\sum\limits_{k \in \mathbb{N}_C} G'_{k,S(C)} P_k + n'_C}{\sum\limits_{j \in \mathbb{N}_M} G'_{j,S(M)} P_j + n'_M}, \tag{4.2}$$

where ν_C and ν_M are, respectively, interference plus noise received at the CUE and the MH. Accordingly, we observe that when $\zeta_{C,M} > 1$, the CUE suffers more severe interference than the MH, and in this case, we refer to ν_C and ν_M as the severe set and the slight set, respectively. By exploiting the concurrent transmission, the CUE can transfer interference from the severe set to the slight set. Note that in order to achieve the maximum EE, interference should be migrated only when the interference intensity index exceeds a certain threshold.

4.2.2 Energy Efficiency of HetNet

In this chapter, we mainly focus on the EE of CUE that implements the interference migration strategy. Since the data of CUE is transmitted by both CUE and MH, the power consumed by MH should be taken into account. Accordingly, the EE of CUE can be defined as the ratio of transmission rate to the total power consumption [20]:

$$\begin{aligned}
\eta_{EE}^{CUE} &= \frac{R}{P} = \frac{R_{CUE} + R_{MH}}{P_{CUE} + P_{MH}} \\
&= \frac{W_c \log_2(1 + \frac{P_{CUE}^{tr}}{v_C}) + W_c \log_2(1 + \frac{P_{MH}^{tr}}{v_M})}{P_{CUE}^{tr} + P_{CM}^{tr} + P_{MH}^{tr} + P_{CUE}^{cst} + P_{MH}^{cst}} \left(\frac{bit}{joule}\right),
\end{aligned} \tag{4.3}$$

where R and W_c denotes the rate requirement of a user, and the transmission bandwidth. R_{CUE} and R_{MH} denote the transmission rate of the subflows transmitted by CUE directly and via the two-hop link, respectively. P_{CUE}^{cst} and P_{MH}^{cst} represent the fixed circuit power of CUE and MH. P_{CM}^{tr} is the power consumption of CUE on the WiFi interface. P_{CUE}^{tr} and P_{MH}^{tr} are the power consumed by CUE and MH on cellular interfaces.

With the capability of concurrent transmission, the user can associate with different kinds of base stations and transmit the traffic via these connections simultaneously. We give an example to explain this, as shown in Figure 4.2. When the CUE performs the concurrent transmission, the traffic of the CUE will be split into two subflows. One subflow is transmitted via the CUE direct link, and the other one is transmitted via the two-hop link. We denote by R_{CUE} the data rate of the direct link and R_{MH} the target rate of the two-hop link. Furthermore, for the two-hop link, we define R_1 and R_2 as the data rate of the first hop and the second hop, respectively. Therefore, we have $R_1 = R_2 = R_{MH}$. In other words, the R_{MH} can be expressed by either R_1 or R_2. We denote the power consumed on the first hop and the second hop as P_{CM}^{tr} and P_{MH}^{tr}, respectively. Due to the short distance and interference-free nature of the WiFi link transmission, the P_{CM}^{tr} is much lower than P_{MH}^{tr} [21]. Thus, we ignore P_{CM}^{tr}

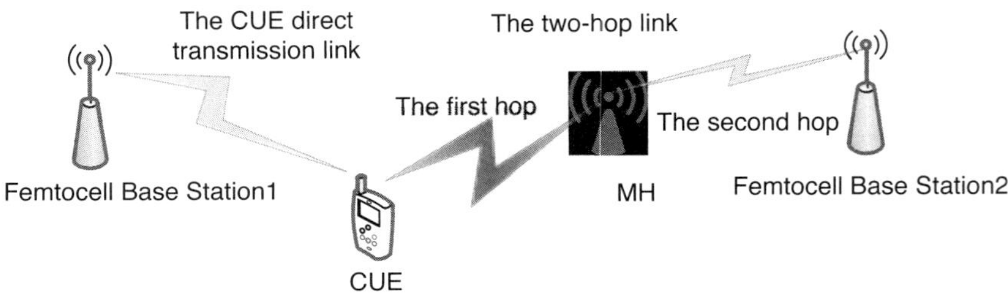

Figure 4.2: An illustration of the concurrent transmission of CUE.

when calculating the energy consumption caused by CUE in the two-hop link. In fact, we have investigated the effect of P_{CM}^{tr} on EE via our simulation results and found that the impact of P_{CM}^{tr} is ignorable. Therefore, the transmission rate of the second hop is given by $R_{MH} = W_c \log_2(1 + \frac{P_{MH}^{tr}}{v_M})$.

4.3 Interference Migration Strategy

In this section, we first illustrate our basic idea to migrate interference through concurrent transmission. Then, we formulate the EE maximization problem and present the solution. Finally, we propose the concurrent transmission interference migration strategy.

Generally speaking, a high data rate requires a high SINR. There are two approaches to improve the SINR: increase the transmit power, or decrease the interference. The former is not a sensible approach, because increasing the transmit power of one UE will cause more severe interference to others who have to increase their own power in order to guarantee their QoS. This power adjustment process is a vicious circle and will cause extremely high power consumption. On the other hand, if we can reduce interference on the transmission of UE, it can decrease the transmit power. Then the total interference in this region becomes less which will lead to a virtuous circle. Obviously, the second approach is more advisable.

Unfortunately, total interference is hardly reduced, especially when traditional power control strategies have been already implemented. However, by executing concurrent transmission, the MH can bridge the different interference neighborhood sets. The CUE can transfer the power from the severe set to slight set by shifting the splitting probabilities. In other words, although total interference is hard to reduce, we can still balance the interference among different spatially separated region by decreasing the interference of one region and correspondingly increasing that of another region. In doing so, the interference is migrated among different regions. More important, the EE can be improved by interference migration even when total interference is not reduced.

4.3.1 EE Maximization via Interference Migration

As we know, the user's data rate is determined by its application type. To achieve minimum power consumption, users prefer to transmit with the minimum rate, which can guarantee their QoS. Hence, no matter what interference management technology has been implemented, the data rate will not be changed. Consequently, maximizing EE is equivalent to minimizing power consumption. Therefore, to maximize EE of CUE, we model the problem of interference migration as

$$\mathbf{P1} : \min P = P_{CUE}^{tr} + P_{MH}^{tr}, \tag{4.4}$$

$$\text{s.t. } R_{CUE} + R_{MH} = W_c \log_2(1 + \frac{P_{CUE}^{tr}}{v_C}) + W_c \log_2(1 + \frac{P_{MH}^{tr}}{v_M}) = R_{req}, \tag{4.5}$$

$$0 \leq P_{CUE}^{tr} \leq P_{CUE}^{\max}, \tag{4.6}$$

$$0 \leq P_{MH}^{tr} \leq P_{MH}^{\max}, \tag{4.7}$$

where R_{req} is the user's data rate requirement, $P_{CUE}^{\max}$ and $P_{MH}^{\max}$ denote the maximum power of CUE and MH, respectively. To satisfy the QoS of CUE, the total transmission rate should be equal to the data rate requirement (i.e., R_{req}). Thus, we have $R_{CUE} + R_{MH} = R_{req}$, where $R_{CUE} + R_{MH}$ is the total transmission rate of CUE. In other words, (4.5) denote that the QoS of CUE should be satisfied. Equations (4.6) and (4.7) represent that the power of CUE and MH must be non-negative and less than their maximum power.

To solve **P1**, we first derive the solution without constraints (4.6) and (4.7), which will be investigated in the next section. Through rearrangement of (4.5), **P1** becomes

$$P_{CUE}^{tr} = \frac{v_C v_M 2^{\frac{R_{req}}{W_c}}}{v_M + P_{MH}^{tr}} - v_C. \tag{4.8}$$

Substituting (4.8) into **P1**, we have

$$\min P = \frac{v_C v_M 2^{\frac{R_{req}}{W_c}}}{v_M + P_{MH}^{tr}} - v_C + P_{MH}^{tr}. \tag{4.9}$$

Let P' denote the first derivative of P with respect to P_{MH}^{tr}. The optimal power of CUE and MH for cellular interface, which is expressed as (4.11) and (4.12), can be derived by

$$P' = \frac{-v_C v_M 2^{\frac{R_{req}}{W_c}}}{(v_M + P_{MH}^{tr})^2} + 1 = 0. \tag{4.10}$$

$$P_{MH}^* = \sqrt{v_C v_M 2^{\frac{R_{req}}{W_c}}} - v_M \tag{4.11}$$

$$P_{CUE}^* = \sqrt{v_C v_M 2^{\frac{R_{req}}{W_c}}} - v_c \tag{4.12}$$

According to (4.11) and (4.12), we can determine the threshold beyond which interference should be migrated.

In the following, we first give the condition to migrate the interference and then show proof of the conclusion.

Interference Migration Condition: The interference should be migrated if and only if the data rate requirement of CUE is larger than $W_c |\log_2(\zeta_{C,M})|$.

This can be proved by the following statement. Since the power is nonnegative, we assume that $\zeta_{C,M} \leq 1$, i.e., $v_C \leq v_M$. According to (4.11) and (4.12), we get $P^*_{CUE} \geq P^*_{MH}$. So we should only guarantee $P^*_{MH} \geq 0$. By setting $P^*_{MH} \geq 0$, we can derive the threshold of R_{req} as:

$$R_{req} \geq W_c \log_2(v_M/v_C) = W_c \log_2(\zeta_{M,C}). \tag{4.13}$$

In contrast, for $\zeta_{C,M} > 1$, i.e., $v_C > v_M$, a similar result can be derived

$$R_{req} \geq W_c \log_2(v_C/v_M) = W_c \log_2(\zeta_{M,C})^{-1}. \tag{4.14}$$

This indicates that whether the interference should be migrated or not depends on the interference intensity index and user's data rate requirement.

4.3.2   Interference Migration via Concurrent Transmission

In this section, we summarize the major phases of the interference migration strategy, which includes the following three phases: (a) measure the interference of v_C directly and receive that of v_C, which is measured by MH; (b) calculate the optimal splitting probabilities based on the R_{req}, v_C and v_M; (c) split the traffic into subflows and distribute them to different interfaces with the calculated probabilities.

The core of the interference migration strategy is phase (b), which will be explained in detail as follows under the condition $\zeta_{C,M} \leq 1$. We consider that the maximum power of CUE can satisfy the user's requirement with the help of admission control schemes. In addition, the power consumption of CUE for interference migration strategy is no more than that for single CUE transmission strategy. Therefore, the constraint $P^{tr}_{CUE} \leq P^{max}_{CUE}$ can be satisfied, and only P^{max}_{MH} should be considered.

Denote $\Psi = \{\varphi_C, \varphi_M\}$ as the set of splitting probabilities of CUE, where φ_C and φ_M represent the probabilities for cellular and WiFi interfaces, respectively. Obviously, $\varphi_C + \varphi_M = 1$. According to the interference migration condition derived in the last section, phase (b) can be divided into the following three cases:

1. $R_{req} \leq W_c \log_2(\zeta_{M,C})$

Traffic will be transmitted by CUE via cellular interface directly, and concurrent transmission will not be executed. Therefore, $\varphi_C = 1$ and $\varphi_M = 0$.

2. $R_{req} > W_c \log_2(\zeta_{M,C})$ and $P^*_{MH} \leq P^{max}_{MH}$

In this condition, concurrent transmission should be executed. The transmit power of CUE and MH should be set according to (4.11) and (4.12), respectively.

Then, the Ψ can be calculated by

$$\begin{cases} \varphi_C = W_c \log_2(1 + \frac{P^*_{CUE}}{v_C})/R_{req} \\ \varphi_M = W_c \log_2(1 + \frac{P^*_{MH}}{v_M})/R_{req}. \end{cases} \qquad (4.15)$$

3. $R_{req} > W_c \log_2(\zeta_{M,C})$ and $P^*_{MH} > P^{\max}_{MH}$

To guarantee constraint (4.7), the power of MH should be modified to $p^{\max}_{MH}$. In this condition, Ψ is expressed as

$$\begin{cases} \varphi_C = 1 - \varphi_M \\ \varphi_M = W_c \log_2(1 + \frac{P^{\max}_{MH}}{v_M})/R_{req}. \end{cases} \qquad (4.16)$$

4.4 Performance Evaluation for Single RAT

In this section, we will evaluate the effects of the proposed strategy via simulations. In the simulation, we consider a HetNet, in which there is one macrocell user equipment and one femtocell user equipment belonging to set $\mathbb{N}_C$ and $\mathbb{N}_M$, respectively (i.e., $\mathbb{N}_C = \{MUE\}$ and $\mathbb{N}_M = \{FUE\}$). The bandwidth of the component carrier is 180 KHz, which is also the bandwidth of a PRB for LTE. The channels are frequency-flat slow fading Rayleigh channels. We first show the feasibility and effectiveness of interference migration by concurrent transmission. We assume that the data rate requirements of the macrocell user equipment, CUE, and femtocell user equipment are 1.4 Mbps, 700 Kbps and 1.2 Mbps, respectively.

In Figure 4.3, we present the interference of $\mathbb{N}_C$ and $\mathbb{N}_M$ as a function of different splitting probability of WiFi interface (i.e., φ_M). It can be seen that the interference gradually migrates from $\mathbb{N}_C$ to $\mathbb{N}_M$ as φ_M increases. We observe that even though the total interference may not be mitigated, the interference can still be migrated among different sets through concurrent transmission.

Figure 4.4 verifies the optimality of the interference migration strategy, which shows the power consumption with different φ_M. The Total curve represents the sum power of CUE and MH. The $\mathbb{N}_C$ suffers more severe interference than $\mathbb{N}_M$ because of the higher data rate requirement of macrocell user equipment. A great deal of energy can be saved by transferring part of the interference from $\mathbb{N}_C$ to $\mathbb{N}_M$ through concurrent transmission. In particular, the minimum power can be obtained by adopting optimal splitting probabilities, which is consistent with our analysis.

Figure 4.5 shows the EE of different UEs with interference migration. For optimal splitting probabilities, the EE of CUE and macrocell user equipment is improved. More important, although EE of femtocell's user equipment is degraded slightly, the total EE of all UEs is still improved by interference migration. This because the slight interference region can tolerate part of the extra interference that is migrated from the severe interference region.

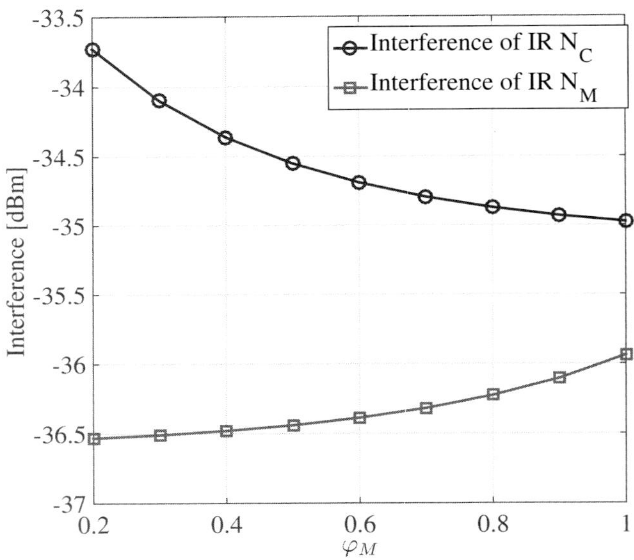

Figure 4.3: Interference of $\mathbb{N}_C$ and $\mathbb{N}_M$ as a function of the splitting probability of WiFi interface φ_M.

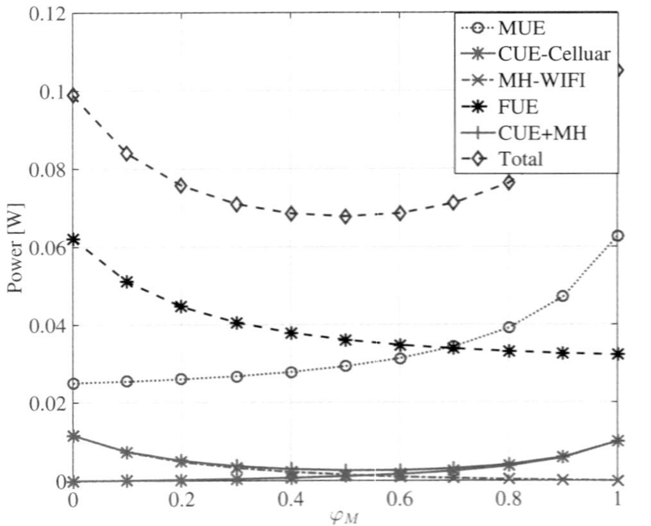

Figure 4.4: Power performance as a function of the splitting probability of WiFi interface φ_M.

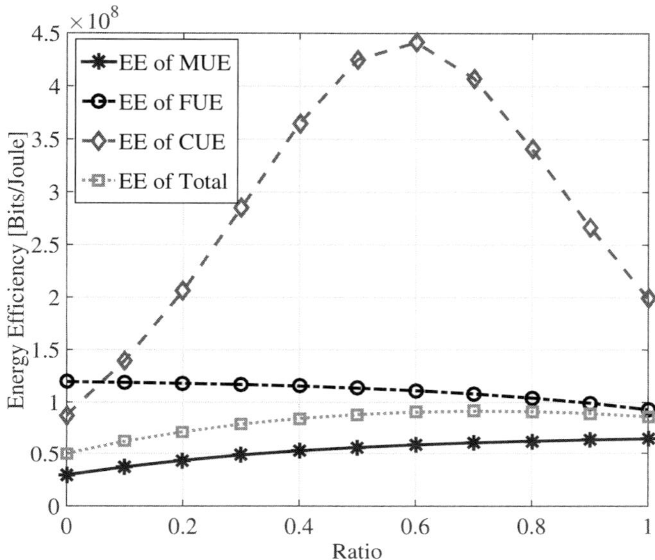

Figure 4.5: EE performance of different UE with interference migration.

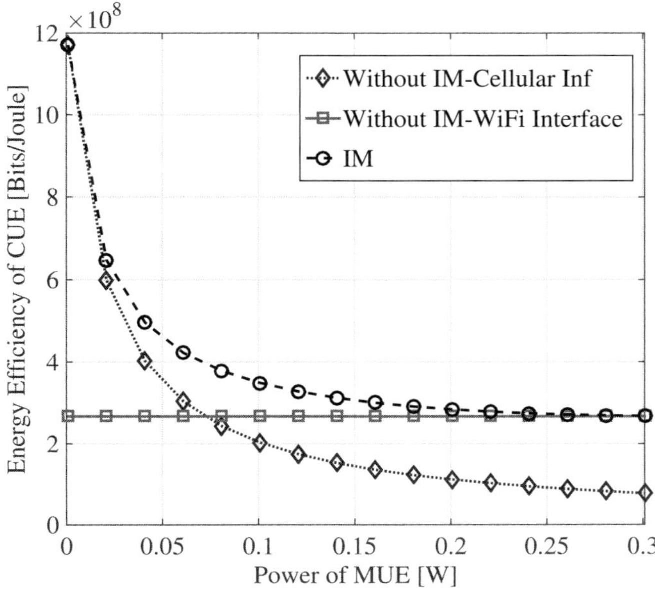

Figure 4.6: EE performance of macrocell user equipment as a function of transmit power.

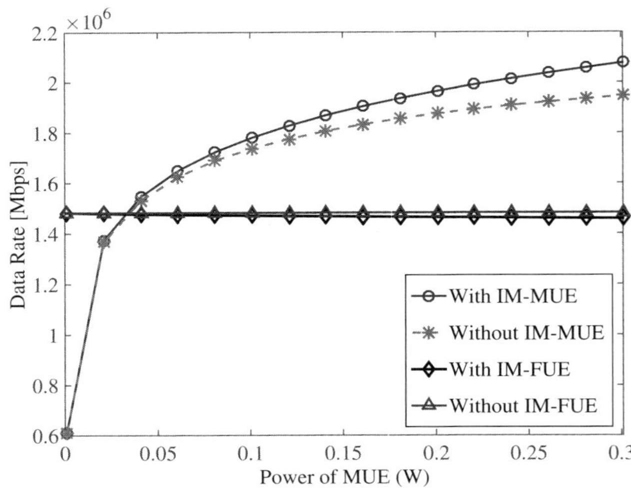

Figure 4.7: Throughput performance of the proposed interference migration strategy.

In Figures 4.6 and 4.7, we show the influence of the interference of $\mathbb{N}_C$ and compare the performance of interference migration with the single-link transmission strategies. In this simulation, we keep the data rate of CUE at 350Kbps and the femtocell user equipment's power at 30 mW. The horizontal axis is the power of macrocell user equipment, which changes from 1mW to 400 mW. The macrocell user equipment dominates the SINR of CUE since it is the only UE in $\mathbb{N}_C$.

Figure 4.6 shows the EE of CUE with different strategies. The black curve represents the EE performance of the interference migration strategy. The dashed line with the diamond marker and the solid line with the square marker are that of single-link transmission without interference migration via cellular and WiFi interfaces, respectively. The cellular interface of CUE should be used exclusively when $W_c \log_2(\psi_{M,C}) \geq R_{req}$. With the rise of the macrocell user equipment's power, the interference should be migrated to the relatively slight set $\mathbb{N}_M$ when $W_c \log_2(\psi_{M,C}) < R_{req}$. As a result, the EE performance of interference migration is better than that of the single transmission strategies. With a further increase of the macrocell user equipment's power, all traffic will be transmitted by MH when $W_c \log_2(\psi_{M,C})^{-1} \geq R_{req}$. It can be seen that the interference migration strategy dominates the single-link transmission strategies for all interference conditions.

Figure 4.7 exhibits the network throughput performance of the proposed interference migration strategy. We compared the throughput performance of macrocell and femtocell user equipment with and without interference migration strategy. It can be seen that with the increase of the macrocell user equipment's

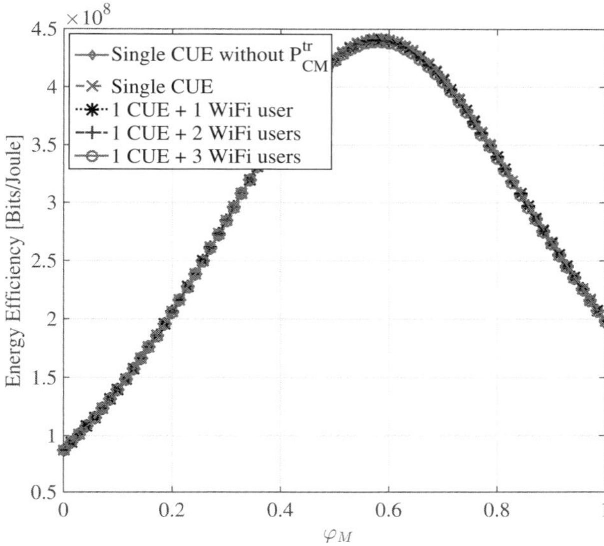

Figure 4.8: The impact of interference in the WiFi network as a function of the splitting probability of WiFi interface φ_M.

power, the throughput of the macrocell user equipment with the interference migration strategy is higher than that without the interference migration strategy remarkably. This can be explained by the fact that the macrocell user equipment locates within the interference-severe region. When the interference migration strategy has been implemented, interference will be transferred out from this region via the CUE. The interference in this region will be reduced. Therefore, transmission performance, such as throughput, will be improved. More important, the heavier interference is in the severe region, the more gain we can obtain.

On the other hand, as we described above, the aim of interference migration strategy is not to reduce total interference, but to transfer interference from one region to another region. Accordingly, as interference reduces in the severe region, interference in the slight region will be increased. However, although the CUE transfers parts of the interference into the slight region, the throughput of femtocell's user equipment, which located in the slight region, almost does not decrease. This phenomenon means that the slight interference region can tolerate some extra interference. Therefore, the proposed interference migration strategy can improve the performance of UE in the severe interference region without causing degradation to the UE in the interference-slight region. In particular, the intersecting line implies that the two regions almost suffer the same interference. This means that it is not necessary to implement interference migration.

In the above analysis, we ignore the transmission power of the WiFi interface

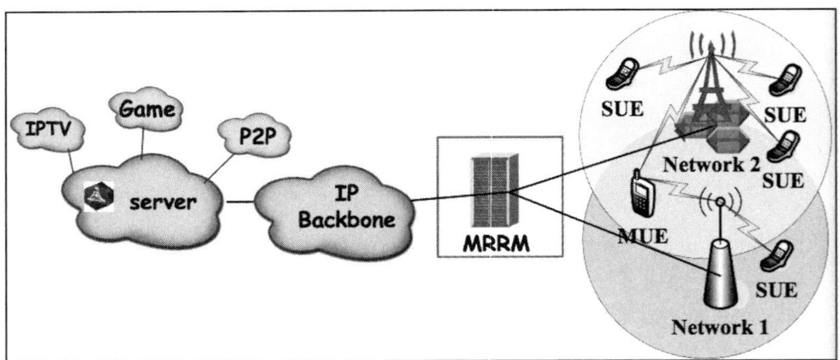

Figure 4.9: HetNet scenario with K multiple RATs.

(i.e., P^{tr}_{CM}) to obtain some insights into this system. In the simulation of Figure 4.8, we consider the influence of P^{tr}_{CM} and investigate the EE performance of CUE when 2-4 WiFi users exist in the HetNet. It can be seen that, due to the short-distance of the WiFi link transmission, the influence of P^{tr}_{CM} on the EE performance is very limited. Therefore, the P^{tr}_{CM} can be neglected in the analysis, while the conclusion dawned above still stand. In fact, if we consider the influence of P^{tr}_{CM} on EE performance, the nonuniform interference distribution phenomenon still exists, and the interference migration can obtain the distribution diversity gain.

4.5   HetNet Model with Multiple RATs

A HetNet may also consist of multiple RAT, such as WiFi, LTE, etc. In the HetNet with multiple RAT, to perform interference mitigation via concurrent transmission, a multimode user equipment (MUE) is required that can obtain multiaccess gain by simultaneously combining transmissions over several RATs by network cooperation [22]. In the following parts, we will show how to maximize the network EE by combining transmissions from multiple RATs.

As shown in Figure 4.9, we consider a HetNet with K ($K \geq 2$) overlapped RATs. Each RAT is entirely independent, and no interference is considered among them. There are two kinds of UEs in our network: the single-mode UE (SUE) accessing with only one RAT and the MUE which can establish concurrent transmission (CT) via different RATs simultaneously. A function entity called MRRM, connected with a CN, is used to manage different RATs and executes flow splitting strategy [23]. In this letter, we only consider the downlink traffic flow. However, a similar consequence can be found in the uplink scenarios.

For the energy efficiency model, we consider a point-to-point transmission in AWGN channel, the data rate can be found by the well-known Shannon formula as

$$R \leq B \log_2 \left(1 + \frac{Pg}{N_0 B}\right), \tag{4.17}$$

where $N_0/2$, B, g and P are the power spectral density, the system bandwidth, the channel power gain from the transmitter to receiver, and the transmit power of the sender, respectively. Meanwhile, the effect of channel fading on data rate can also be considered in (4.17). (The channel power gain g is expressed as $g = \|h_{tr}\|^2$, where h_{tr} denotes the channel gain from the transmitter to the receiver. Therefore, the channel fading can be reflected by the change of channel power gain g.) Let η_{EE} be the system EE, which is defined as the number of bits that the network can deliver per joule of energy. Based on (4.17), η_{EE} can be expressed as

$$\eta_{EE} = \frac{R}{P_{BS}} = \frac{R}{\left(2^{R/B} - 1\right)\left(N_0 B/g\right) + P_{cst}} \; [Bits/Joule]$$

where P_{BS} is the total power consumed by the BS, and denotes the fixed power consumption due to transporting and processing units.

4.6 EE Maximization via Concurrent Transmission

In this section, we first explore the intrinsic relationship between energy consumption and transmission rate, and then propose a concurrent transmission scheme, MECT, to maximize the system EE.

4.6.1 Problem Formulation

Let $\mathbb{R} = \{R_1, R_2, ..., R_K\}$ denote the vector of flows carried by the corresponding RATs that represents the downlink data rate from RATs to MUE. R_i means the traffic will not transmit via RAT_i. Due to the existence of SUE, all BSs must be in a working state, and the influence of P_{cst} is the same for all BSs no matter whether they are splitting traffic or not. Thus, in the following discussions, we neglect the value of P_{cst}. Without loss of generality, we use the upper bound of transmission capacity as predicted by the Shannon theorem, and the maximum EE can be calculated as follows:

$$\max \eta_{EE} = \max \frac{\sum_{i=1}^{i=K} R_i}{\sum_{i=1}^{i=K} P_{BS_i}(R_i)}$$

$$= \max \frac{R_{req}}{\sum_{i=1}^{i=K}\left(2^{R_i/B_i} - 1\right)\left(N_0 B_i/g_i\right)}, \tag{4.18}$$

$$s.t. \; R_{req} = \sum_{i=1}^{K} R_i, \tag{4.19}$$

$$0 \leq R_i \leq R_{iMAX}, \tag{4.20}$$

where P_{BS_i} denotes power consumption of the ith BS, and B_i denotes the available resources of the ith BS which can be utilized by an MUE. R_{req} is the data requirement of users. Equation (4.20) represents that the data rate of the i-th network must be less than its maximum transmit capacity (in this chapter, we assume that the maximum data rate of each RAT can satisfy the QoS of the user). Because the data rate requirement of the user is irrelevant with different energy saving strategies, (4.18) is equivalent to (4.21).

Because the data rate requirement of the user R_{req} is irrelevant with different energy saving strategies, (4.18) is equivalent to (4.21).

$$\min \sum_{i=1}^{K} P_{BS_i}(R_i) = \sum_{i=1}^{K} \left(\left(2^{R_i/B_i} - 1 \right) (N_0 B_i / g_i) \right). \tag{4.21}$$

As shown in (4.21), the target function is the sum of a set of exponential functions, and therefore this function is convex with respect to $\mathbb{R}$. Thus, we can easily derive the optimal rate R_i^* of each RAT utilizing the method introduced in [24] under a given available resources (such as available bandwidth B_i).

4.6.2 Existence of Feasible Solutions

In this part, we will derive a feasible solution for the problem described above in terms of the first derivatives P_{BS_i}'. Let $P(\mathbb{R})$ and $\partial P(\mathbb{R})/\partial R_i$ denote the power consumption and its partial derivative of the system, respectively.

Let $\mathbb{R}^* = \{R_i^* | i = 1, 2, ..., K\}$ be the allocated rate of the optimal MECT strategy. For any $R_i^* > 0$, if we shift a small amount of flow $\delta > 0$ from RAT_i to another RAT_j, the total power consumption of all RATs must not reduce. Otherwise, the optimality of $\mathbb{R}^*$ would be violated. The change of the power caused by the shift can be expressed as

$$\delta \frac{\partial P(\mathbb{R}^*)}{\partial R_j} - \delta \frac{\partial P(\mathbb{R}^*)}{\partial R_i} \geq 0. \tag{4.22}$$

For the same transmission rate, without loss of generality, we can sort K different BSs as $P_{BS_1}(x) \leq L \leq P_{BS_i}(x) \leq L \leq P_{BS_K}(x)$ by the power consumption. Hence, we have $R_1^* \geq L \geq R_i^* \geq L \geq R_K^* \geq 0$, and there will be K possibilities of the optimal $\mathbb{R}^*$. Namely, for the ith possibility, we have $\begin{cases} R_j^* > 0, & 1 \leq j \leq i \\ R_j^* = 0, & i < j \leq K \end{cases}$. Let R_b^i be the threshold of the ith BS, we have $R_b^i < +\infty$, $i \in [1, K-1]$, and $R_b^K = +\infty$ (due to the Kth BS is the worst). If $R_i^* > R_b^i$, $i \in [1, K-1]$, the system will enter the $(i+1)$-th possibility. Furthermore, the optimal data rate for each RAT (in the (i+1)st possibility) can be calculated by (4.23), which is derived by combing (4.19) and (4.22).

$$\begin{cases} \begin{bmatrix} 1 & -1 & 0 & L & 0 & 0 \\ 0 & 1 & -1 & L & 0 & 0 \\ 0 & 0 & 1 & L & 0 & 0 \\ M & M & M & O & M & M \\ 0 & 0 & 0 & L & 1 & -1 \end{bmatrix} \begin{bmatrix} \frac{\partial P(\mathbb{R}^*)}{\partial R_1} \\ \frac{\partial P(\mathbb{R}^*)}{\partial R_2} \\ M \\ \frac{\partial P(\mathbb{R}^*)}{\partial R_K} \end{bmatrix} = [0] \\ R_{req} = \sum_{i=1}^{K} R_i \end{cases} \tag{4.23}$$

In what follows, a HetNet scenario with two different RATs is considered, but the general properties and performance measures hold for $K \geq 2$. The power consumed by both RATs can be represented as

$$P(\mathbb{R}) = P_{BS_1}(R_1) + P_{BS_2}(R_2). \tag{4.24}$$

Because of $R_1^* \geq R_2^*$, then there are only two possibilities as follows:

- $R_1^* = R_{req}$ and $R_2^* = 0$

With this condition, we have the following conclusions: If and only if the data rate requirement of an MUE is larger than $B_1 \log_2 (g_2/g_1)$, the concurrent transmission should be started.

This can be proved by deriving the derivative of power consumption with respect to data rate. According to (4.23), the following inequality can be obtained as

$$dP(0)/dR_2 \geq dP(R_{req})/dR_1. \tag{4.25}$$

Substituting $P'(R) = (N_0/g) 2^{R/B} \ln 2$ into (4.25), we can rewrite the inequality as

$$(N_0/g_2) \ln 2 \geq (N_0/g_1) 2^{\frac{R_{req}}{B_1}} \ln 2, \tag{4.26}$$

$$R_{req} \leq B_1 \log_2 (g_1/g_2). \tag{4.27}$$

Simplifying (4.26), we can get the threshold values $R_b^1 = R_b = B_1 \log_2 (g_1/g_2)$ at which the CT should be started, and $R_b^2 = +\infty$. In other words, the CT will reduce the energy consumption only when the data rate requested by an MUE is larger than $B_1 \log_2 (g_1/g_2)$.

- $R_1^* > 0$ and $R_2^* > 0$

In this case, the CT should be implemented, and the condition for minimizing the energy consumption implies the following relationship;

$$dP(R_1^*)/dR_1 = dP(R_2^*)/dR_2, \tag{4.28}$$

which leads to $(N_0/g_1) 2^{\frac{R_1^*}{B_1}} \ln 2 = (N_0/g_2) 2^{\frac{R_2^*}{B_2}} \ln 2$; rewriting this equation we have $B_1 R_2^* - B_2 R_1^* = B_1 B_2 \log_2 (g_2/g_1)$.

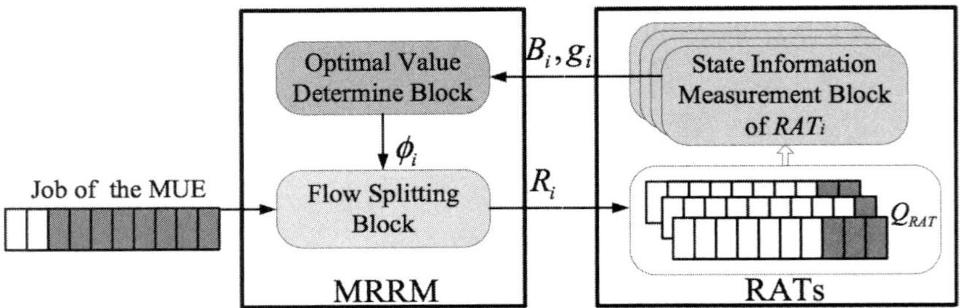

Figure 4.10: MECT strategy block diagram.

Combined with the condition $R_1^* + R_2^* = R_{req}$, we can derive the optimal value of R_1^* and R_2^* as (4.29).

$$\begin{cases} R_1^* = \dfrac{B_1 R_{req} - B_1 B_2 \log_2 \frac{g_2}{g_1}}{B_1 + B_2} \\ R_2^* = \dfrac{B_2 R_{req} - B_1 B_2 \log_2 \frac{g_2}{g_1}}{B_1 + B_2} \end{cases} \tag{4.29}$$

It can be seen that RAT_1 will be used exclusively when the $R_{req} \leq B_1 log_2 (g_1/g_2)$. Otherwise, the CT should be adopted, and RAT_2 will be also utilized. As R_{req} continues to increase, the flows on all the RATs augment correspondingly. However, the optimal rate of each RAT should obey the restriction of (4.28). And the additional RATs are used to avoid overloading the advanced RAT.

4.6.3 MECT Scheme

There are three important entities of MECT scheme, as shown in Figure 4.10, state information measurement (SIM), optimal value determine (OVD), and flow splitting (FS). The SIM module in each RAT estimates the RAT 's information (such as the available bandwidth B_i, channel power gain g_i, etc.) that can be utilized by the MUE and reports this information to the OVD located in MRRM. The OVD derives the optimal data rate of each RAT and calculates the optimal splitting probability by (4.30), in which the value of R_i is derived via (4.18)-(4.20). Finally, the FS entity distributes the arriving packets to the queue of corresponding RAT according to the probability ϕ_i.

$$\phi_i = R_i/R_{req} \tag{4.30}$$

4.7 Comparison of Solutions

The system is designed as follows: the available bandwidths for the MUE are $B_1 = B_2 = 100\text{KHz}$, and the corresponding channel power gains are $g_1 = 0.008$, $g_2 = 0.004$ and $N_0 = 10^{-9}\text{W/Hz}$. With these parameters, the threshold value R_b is 0.1 Mbps at which the CT is adopted.

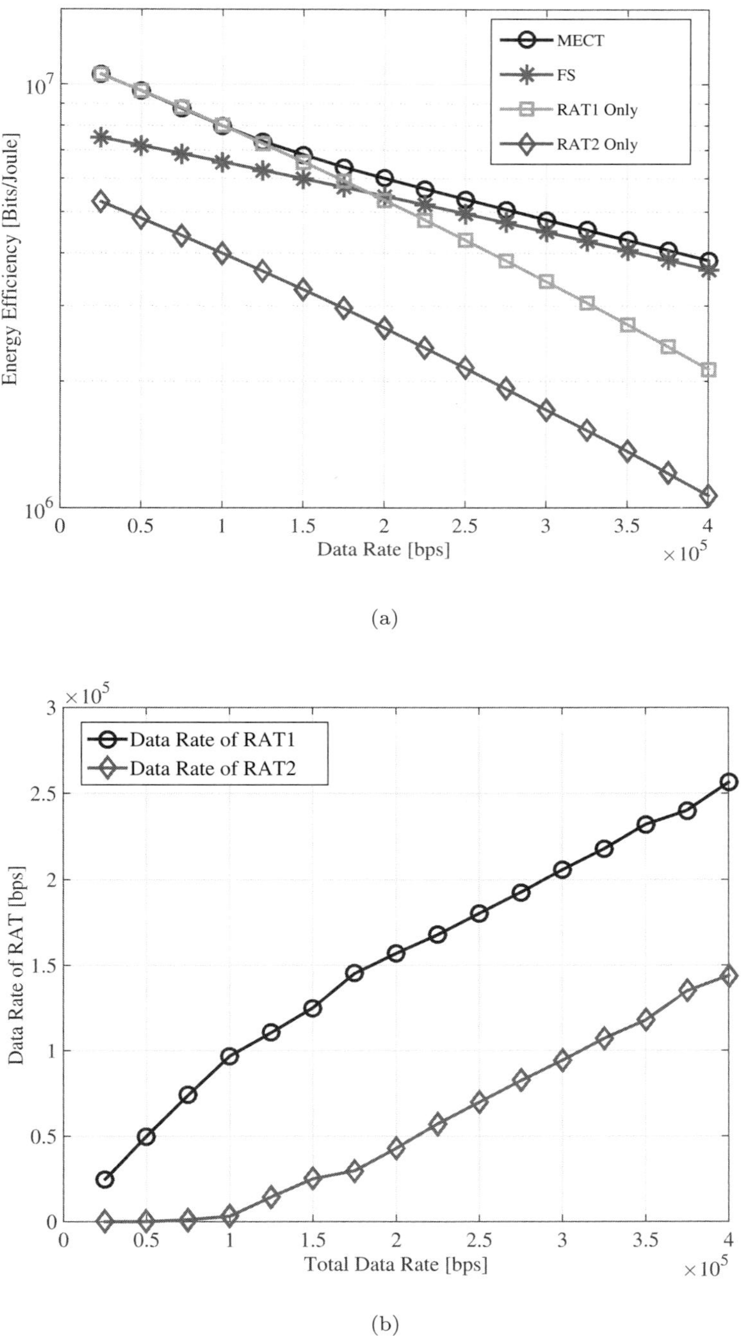

(a)

(b)

Figure 4.11: Energy efficiency performance (a) and data rate of RATs (b) as a function of data rates.

We compare the MECT with single RAT transmission and another well-known CT strategy, FS [25]. The single RAT transmission that transmits whole job via only one RAT can be regarded as the sleep mechanisms (SM) strategy proposed in [26]. The FS is designed to achieve maximum throughput of whole system by splitting the flow with fixed probability corresponding to the throughput of each RAT.

Figure 4.11(a) shows the EE of HetNets for the MUE. The horizontal and vertical axes represent the MUE's data rate requisition and EE, respectively. It's clear that the EE of a single RAT decreases much quicker than that of the MECT with the growth of a data rate. That means the MECT can save significant power especially in a high data rate. It's worth noting that RAT_1 will be used exclusively when $R_{req} > R_b = 0.1\text{Mbps}$. We can also see that the performance gap between the MECT and the FS becomes smaller with the increase of the data rate. The reason is because the performance for CT is determined by the transmission rate vector $\mathbb{R}$, which is dependent on the ϕ_i. Within certain states of HetNets, ϕ_i^{FS} will be a constant. While ϕ_i^{MECT} will change with the variation of R_{req} and maintain the optimal ϕ_i. With the further increase of R_{req}, ϕ_i^{FS} will be close to ϕ_i^{MECT} and make the performance gap between the MECT and the FS smaller. However, the MECT is dominant over the FS.

The optimal rate of each RAT is shown in Figure 4.11(b). RAT_1 will carry the entire traffic at the beginning. With the increase of the data rate, traffic flow will be distributed into both of the RATs based on the probability calculated by (4.30). In Figure 4.12, we will show the influence of channel power gain. We keep R_{req} at 0.1 Mbps and let the channel power gain of RAT_1 change from 0.00525 to 0.00875 gradually. It's clear that the MECT always has the best performance when the channel power gain changed. It's worth noting that in the low data rate region, the FS has a worse performance than even the single RAT transmission. This means that splitting flow with an improper probability may worsen the performance and waste extra power.

In Figure 4.13, we will show the influence of the available bandwidth. We keep the data rate at 1 Mbps. The horizontal axis is the available bandwidth of RAT_1 that can be used by the MUE. It can be seen that with the increase of available bandwidth, the MECT has a better performance than both the FS and single RAT transmission.

From Figure 4.13, we find that as the available bandwidth increases, the EE performance gap between the MECT and the FS expands. As we know, the splitting probability changes with the variation of available bandwidth and the optimal data rate vector $\mathbb{R}^*$ of the MECT is derived by the constrained optimization problem (4.18)-(4.20). Thus, the splitting probability of the MECT can utilize the resources of the HetNets most efficiently and obtain the optimal EE performance, while that both ϕ_1^{MECT} and ϕ_1^{FS} are small when the bandwidth of RAT_1 is limited. Thus, the MECT and the FS have very a close performance in the low-bandwidth region.

Figure 4.14 shows the performance of MECT in a four-RATs HetNet sce-

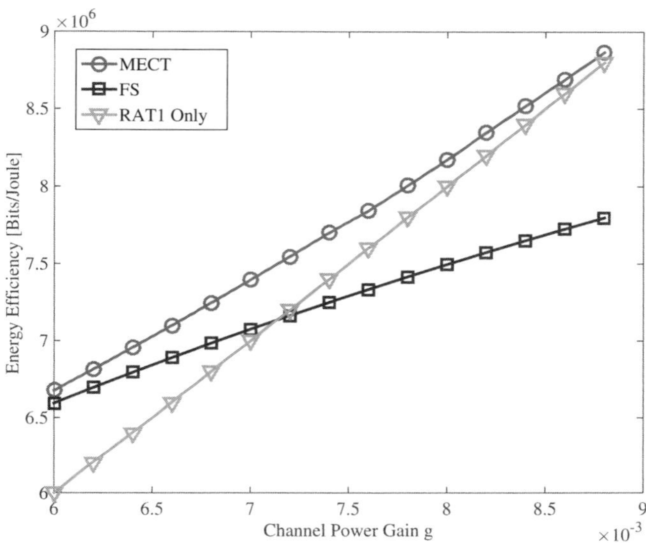

Figure 4.12: Comparison of different strategies with different channel power gain.

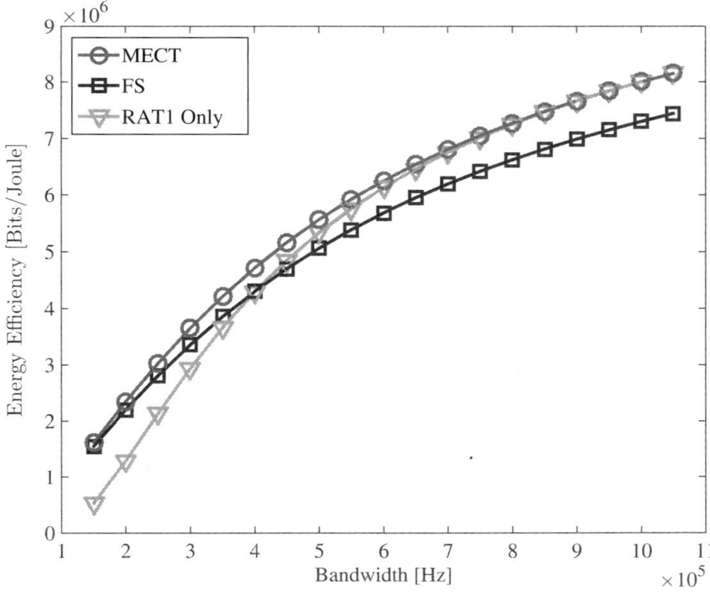

Figure 4.13: Comparison of different strategies with different bandwidth.

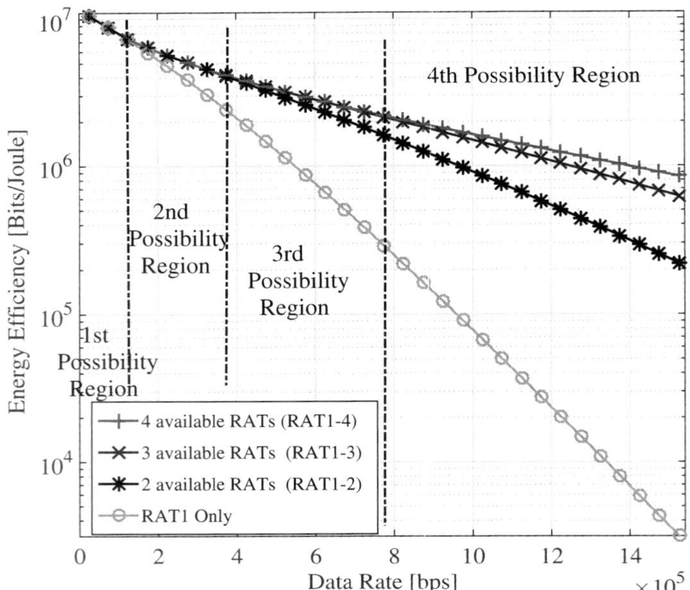

Figure 4.14: The performance of MECT in a multiple RATs network scenario.

nario. The channel power gain of the four RATs are $g_1 = 0.008$, $g_1 = 0.004$, $g_1 = 0.002$, and $g_1 = 0.001$, respectively. The whole transmission process is divided into four possibilities. With the increase of R_{req}, the process will successively enter different possibilities, and the more available RATs we have, the better EE performance of system we can obtain. However, since RATs will be used in the order of their conditions, the rate of EE performance gain will slow down with the increasing number of available RATs.

4.8   Conclusions

In this chapter, we proposed a novel interference management strategy, namely interference migration, to deal with the nonuniform interference distribution phenomenon. Interference migration is the process of transferring interference from one region to another region by implementing concurrent transmissions. We first illustrated the interference migration scheme within a single RAT Het-Net and then extended the concurrent transmission to the multiple RATs Het-Net. Furthermore, we use the energy efficiency to measure the performance of the proposed strategy. The simulation results show that although the total interference is hard to reduce, the proposed strategy can still balance the interference among different spatially separated regions, and improve the energy efficiency performance remarkably.

Furthermore, to obtain some insight into this system, we used the scenario with a single mobile hotspot and single CUE to analyze the property of the proposed strategy. In practical systems, there may exist more than one mobile hotspot and CUE with much more complex channel conditions. By introducing the multiple hotspot selection scheme, the proposed strategy can be easily extended to the single CUE and the multiple mobile hotspots scenario. In contrast, for the multiple CUE scenario, due to the mutual interference and the limited knowledge among different CUEs, the problem becomes more complex and cannot be solved directly with the method described in this chapter. Nonetheless, the nonuniform interference distribution phenomenon still exists, and interference migration should still obtain distribution diversity gain.

REFERENCES

[1] N. Saquib, E. Hossain, L. B. Le, and D. I. Kim, "Interference management in ofdma femtocell networks: issues and approaches," *IEEE Wireless Communications*, vol. 19, no. 3, pp. 86–95, June 2012.

[2] Y. Meng, L. I. Jiandong, and L. I. Hongyan, "Efficient resource allocation scheme for multi-service based on interference mitigation in lte-advanced networks," *Science China Information Sciences*, vol. 57, no. 8, pp. 1–11, 2014.

[3] J. G. Andrews, "Interference cancellation for cellular systems: a contemporary overview," *IEEE Wireless Communications*, vol. 12, no. 2, pp. 19–29, April 2005.

[4] S. P. Weber, J. G. Andrews, X. Yang, and G. de Veciana, "Transmission capacity of wireless ad hoc networks with successive interference cancellation," *IEEE Transactions on Information Theory*, vol. 53, no. 8, pp. 2799–2814, Aug 2007.

[5] L. G. F. Trichard, J. S. Evans, and I. B. Collings, "Large system analysis of linear parallel interference cancellation," in *ICC 2001. IEEE International Conference on Communications. Conference Record (Cat. No.01CH37240)*, vol. 1, Jun 2001, pp. 26–30 vol.1.

[6] T. Zahir, K. Arshad, A. Nakata, and K. Moessner, "Interference management in femtocells," *IEEE Communications Surveys Tutorials*, vol. 15, no. 1, pp. 293–311, First 2013.

[7] V. Chandrasekhar, J. G. Andrews, Z. Shen, T. Muharemovic, and A. Gatherer, "Distributed power control in femtocell-underlay cellular networks," in *GLOBECOM 2009 - 2009 IEEE Global Telecommunications Conference*, Nov 2009, pp. 1–6.

[8] T. S. N. Fan M and M. F, "Interference management in femtocell deployment," in *In Proceedings of 3GPP2 Femto Workshop.*, 2007.

[9] M. Rahman and H. Yanikomeroglu, "Enhancing cell-edge performance: a downlink dynamic interference avoidance scheme with inter-cell coordination," *IEEE Transactions on Wireless Communications*, vol. 9, no. 4, pp. 1414–1425, April 2010.

[10] O. Onireti, F. Heliot, and M. A. Imran, "On the energy efficiency-spectral efficiency trade-off in the uplink of comp system," *IEEE Transactions on Wireless Communications*, vol. 11, no. 2, pp. 556–561, February 2012.

[11] J. T. Louhi, "Energy efficiency of modern cellular base stations," in *INTELEC 07 - 29th International Telecommunications Energy Conference*, Sept 2007, pp. 475–476.

[12] A. Carroll and G. Heiser, "An analysis of power consumption in a smartphone," in *Usenix Conference on Usenix Technical Conference*, 2010, pp. 21–21.

[13] Y. Kim, H. Ko, S. Pack, W. Lee, and X. Shen, "Mobility-aware call admission control algorithm with handoff queue in mobile hotspots," *IEEE Transactions on Vehicular Technology*, vol. 62, no. 8, pp. 3903–3912, Oct 2013.

[14] K. Keshav, V. R. Indukuri, and P. Venkataram, "Energy efficient scheduling in 4g smart phones for mobile hotspot application," in *2012 National Conference on Communications (NCC)*, Feb 2012, pp. 1–5.

[15] Z. Yang, Q. Yang, F. Fu, and K. S. Kwak, "A novel load balancing scheme in lte and wifi coexisted network for ofdma system," in *2013 International Conference on Wireless Communications and Signal Processing*, Oct 2013, pp. 1–5.

[16] S. Dimatteo, P. Hui, B. Han, and V. O. K. Li, "Cellular traffic offloading through wifi networks," in *2011 IEEE Eighth International Conference on Mobile Ad-Hoc and Sensor Systems*, Oct 2011, pp. 192–201.

[17] F. Zhang, W. Zhang, and Q. Ling, "Non-cooperative game for capacity offload," *IEEE Transactions on Wireless Communications*, vol. 11, no. 4, pp. 1565–1575, April 2012.

[18] B. Han, P. Hui, V. S. A. Kumar, M. V. Marathe, J. Shao, and A. Srinivasan, "Mobile data offloading through opportunistic communications and social participation," *IEEE Transactions on Mobile Computing*, vol. 11, no. 5, pp. 821–834, May 2012.

[19] A. Sharma, V. Navda, R. Ramjee, V. N. Padmanabhan, and E. M. Belding, "Cool-tether: energy efficient on-the-fly wifi hot-spots using mobile phones," in *International Conference on Emerging NETWORKING Experiments and Technologies*, 2009, pp. 109–120.

[20] G. Miao, N. Himayat, and G. Y. Li, "Energy-efficient link adaptation in frequency-selective channels," *IEEE Transactions on Communications*, vol. 58, no. 2, pp. 545–554, February 2010.

[21] J. Huang, F. Qian, A. Gerber, Z. M. Mao, S. Sen, and O. Spatscheck, "A close examination of performance and power characteristics of 4g lte networks," in *International Conference on Mobile Systems, Applications, and Services*, 2012, pp. 225–238.

[22] A. Bazzi, G. Pasolini, and O. Andrisano, "Multiradio resource management: Parallel transmission for higher throughput?" *EURASIP Journal on Advances in Signal Processing*, vol. 2008, no. 1, pp. 1–9, 2007.

[23] R. Ferrus, O. Sallent, and R. Agustš[a], *Interworking in heterogeneous wireless networks: Comprehensive framework and future trends.* IEEE Press, 2010.

[24] S. Boyd and L. Vandenberghe, "Convex optimization," *Cambridge University Press*, 2004.

[25] G. J. Hoekstra, R. D. van der Mei, and S. Bhulai, "Optimal job splitting in parallel processor sharing queues," *Stochastic Models*, vol. 28, no. 1, pp. 144–166, 2012.

[26] L. Saker, S. E. Elayoubi, and T. Chahed, "Minimizing energy consumption via sleep mode in green base station," in *IEEE Wireless Communications and NETWORKING Conference, WCNC 2010, Proceedings, Sydney, Australia, 18-21 April,* 2010, pp. 1–6.

Chapter 5

Distributed Resource Allocation for Interference Mitigation

5.1 Introduction

Femtocells are a promising technology to fulfil the explosive demand for high data rate services and the requirement of ubiquitous access [1–3]. To be specific, in contrast to conventional MBSs, femtocell access points (FAPs) are lower-power, short-range, plug-and-play small BSs that are installed and managed by customers in residential areas and small offices. With the help of this novel technology, more users can share the same spectrum resource by accessing different femtocells in different areas. On the other hand, users in poor indoor coverage regions or dead zones can achieve better performance by deploying additional FAPs. Nevertheless, to exploit the benefits promised by femtocells, it is necessary to mitigate both cotier and cross-tier interference, which emerge when the same spectrum is allocated to different cells [4]. In particular, cotier interference refers to interference among femtocells or macrocells, whereas cross-tier interference is interference between macrocells and femtocells.

OFDMA technology can exploit channel variations in both frequency and time domains by dividing the available spectrum into orthogonal subchannels. In OFDMA femtocell networks, dedicated-channel deployment (or orthogonal channel assignment) can be used to cope with cross-tier interference (i.e., femtocells would be allocated a fraction of subchannels while the macrocell would be allocated another fraction [4–6]). Although this strategy is suboptimal from a spectral efficiency standpoint, it is a relatively simple solution to avoid cross-interference and has been considered as one option for eICIC in LTE Release 10 specifications [7].

When dedicated channel deployment is adopted in an OFDMA femtocell network, cross-tier interference is avoided but cotier interference still needs to be considered. Dynamic frequency allocation is a promising technique to address this issue due to its low communication overhead [8]. By constructing an interference graph for a femtocell network, femtocell grouping-based and greedy-based resource allocation strategies were investigated in [9] and [10], respectively. Energy-efficient resource allocation for cognitive radio (CR) femtocell networks was studied in [11], where cognitive BSs sold the spectrum bought from the primary networks to FAPs. This problem was formulated as a Stackelberg game and a gradient based iteration algorithm was devised to achieve the equilibrium state.

In all of the above mentioned works [9–11], central controller is necessary for femtocell management. However, because of the uncertainty in the number and positions of the femtocells, it is not viable to perform a centralized resource allocation [12]. Even if possible, it would require significant signaling overhead and cause congestion in the backhaul network[1] [4]. For this reason, some researchers begin to turn their attentions to distributed interference mitigation in OFDMA femtocells [8, 14–19]. To be specific, dedicated subchannels were allocated to femtocells (or a femtocell user (FU)) in a random manner in [14, 15]. In [8, 16, 17], when choosing from the available subchannels, each FAP or FU just considered the existing interference on each subchannel (i.e., incoming interference), which may result in severe interference within the network and reduced the overall capacity of femtocells. To further improve the overall capacity, not only incoming interference but also outgoing interference (i.e., interference caused to existing femtocells) were considered in [18, 19]. In particular, the proposed autonomous component carrier selection and its improved version could achieve better aggregate performance at the cost of higher signaling overhead.

Therefore, it is necessary to study the issue of distributed subchannel allocation in OFDMA femtocell networks without signaling overhead. In a decentralized network, there are two kinds of conflicts. One is the conflict between different femtocells that is caused by the limitation of spectrum resource and coupling among different femtocells. The other is the conflict between system performance and individual requirement that is mainly introduced by the lack of a central controller. In fact, these two conflicts always make a decentralized network operate at an inefficient point, which is termed the price of anarchy (PoA). In this light, to exploit the benefits promised by the decentralized networks, it is essential to design distributed resource allocation strategies that should fully consider these two conflicts. Fortunately, game theory, which provides a suitable paradigm to analyze the interrelationship between decision makers, can be naturally adopted to deal with the first conflict [20–25]. However, designing globally optimal or even pareto-efficient (pareto-optimal)[2] distributed resource allocation algorithms for a decentralized network is still an open problem [21–23].

---

[1] In a femtocell network, FAPs can transmit limited signaling data over the backhaul network via residential wireline broadband access links (e.g., digital subscriber lines (DSL) [13]).

[2] Generally speaking, it is easy to prove that the global optimal solution is also pareto-optimal, but not vice versa.

As a mathematical tool for analyzing the conflict and cooperation between autonomous agents, game theory has been widely used to design interference mitigation schemes in multicell OFDMA systems [26–28]. We would like to note that these distributed strategies can be easily extended to OFDMA femtocells. Specifically, rate maximization with a power pricing game was formulated in [26] and a NE could be achieved with the proposed algorithm in the case when all cells operated in full load (i.e., all subchannels were utilized). However, in general cases, BSs would choose subchannels according to traffic load and then there is no guarantee that the NE for NRMG always exists [27]. As pointed out in [28], due to the nonexistence of NEs, it is a challenge to devise a game based learning algorithm to achieve stable solutions. To get around this difficulty, both [27] and [28] formulated the distributed subchannel allocation as a potential game, where there is a wide class of learning algorithms converging to a pure NE (e.g., gradient play, fictitions play, and joint strategy fictitious play [29]).

However, there are some inherent shortcomings in the formulated potential game, that would affect the performance of the proposed algorithm. For instance, to formulate a potential game, the corresponding utility function should be properly designed. As a result, there may be no obvious relationship between the utility and capacity. In other words, an increase in total utility is not equivalent to a higher overall capacity [27, 28]. As well, the efficiency of the achieved NE cannot be guaranteed and the information exchanged among players requires additional overhead.

In this chapter, we first formulate the subchannel allocation problem as a noncooperative rate maximization game (NRMG) where the utility of each femtocell is its capacity. However, it has been proven that there is no guarantee that the Nash equilibrium (NE) for NRMG always exists, which makes it more difficult to design distributed strategies to achieve stable and efficient solutions. To get around this difficulty, previous studies generally change the player's utility function, which may cause the loss of system capacity. However, motivated by the studies of the utility-based leaning model that is appropriate for studying multiagent systems [30–33], we have introduced an additional state space in NRMG, and then proposed a utility-based distributed subchannel allocation algorithm (UDSA). In addition, to avoid the high communication overhead, we focus on a network where there is no information exchange among different users (i.e., no common control channel (CCC) is introduced). We note that this consideration makes the scenario more practical but on the other hand brings us more difficulties in designing efficient resource allocation strategies [21, 34–38].

Then, we consider both subchannel allocation and power control and formulate this problem as a noncooperative transmission control game (NTCG). To overcome the lack of communication between different users, a utility-based learning approach is adopted and a utility-based transmission control (UTC) algorithm is developed, with which each user can configure its operational parameters just by measuring local interference. More important, although there is no guarantee that the NE for NTCG always exists, it is proven that the femtocells could operate at a global optimal point by implementing UTC. Finally, simulation results verify the validity of our analysis and demonstrate that the performance of our algorithm (convergence speed, achieved overall utility, etc.) are better than that of the existing distributed algorithms.

5.2   Distributed Subchannel Allocation

5.2.1   System Model and Problem Formulation

Let's consider a femtocell network, where FAPs are deployed in random locations and configured in closed access mode (CAM)[3]. In particular, there are in total N femtocells sharing the dedicated spectrum consisting of K orthogonal subchannels, each of which has bandwidth B_0. Therefore, the cross-tier interference in this network is avoided. For a given slot, if different FAPs (or FUs) transmit on the same subchannel, downlink (or uplink) cotier interference occurs. Actually, in each femtocell, multiple femto users can be scheduled by using OFDMA. For analysis simplicity, we consider that there is only one FU communicating with the FAP in each time slot,[4] which has been assumed in previous studies due to the opportunistic scheduling operation [11, 39]. In this work, we mainly study interference mitigation for downlink communication. However, under the above assumptions, the proposed scheme is also suitable for the uplink case. Without loss of generality, we label the FU subscribing to FAP n by FU n. As well, let us assume that each FAP n chooses K_n subchannels to be assigned for the data transmission to FU n, where $0 < K_n < K, \forall n \in \mathcal{N}$.[5]

Let us denote the channel gain matrix by $\mathbf{G} \in \mathbb{R}^{N \times N \times K}$, where element $g_{n,m}^k$ represents the channel gain between FAP n and FU m on subchannel k. For analysis simplicity, we just consider path loss and shadowing fading. In particular, let $d_{n,m}$ denote the distance between FAP n and FU m, and $g_{n,m}^k$ in dB is [40]

$$\begin{cases} -28 - 35\log_{10}\left(d_{n,m}\right) - \psi, & n = m \\ -38.5 - 20\log_{10}\left(d_{n,m}\right) - L_{wall} - \psi, & n \neq m \end{cases} \quad (5.1)$$

where L_{wall} and ψ are the wall penetration loss and Gauss-distributed random variable with mean zero and variance σ_ψ^2, respectively. It is assumed that each FAP n only knows the channel gains between itself and its subscriber (i.e., $g_{n,n}^k$, $k \in \{1, 2, \cdots, K\}$). Compared to the assumptions in [27, 28], we dispense with a large signaling overhead to exchange CSI between any two different femtocells; that is, $g_{m,n}^k$ and $g_{n,m}^k$ where $\forall n, m \in \{1, 2, \cdots, N\}, m \neq n$, and $\forall k \in \{1, 2, \cdots, K\}$.

Let $\mathcal{S}_n$ be the set of subchannels assigned to FU n; that is, $\mathcal{S}_n = \{s_1, s_2, \cdots, s_{K_n}\}$, where $s_k \in \{1, 2, \cdots K\}$ is the index of the kth subchannel allocated to

---

[3]Three access modes have been defined for a femtocell network: open access mode (OAM), CAM, and hybrid access mode, respectively [4].

[4]It is worth noting that the proposed algorithm can be extended to multiple users per time slot by enlarging the policy set, but the computation complexity would be higher.

[5]Note that when $K_n = 0$ or $K_n = N$, there is no subchannel selection for FAP n. In this case, the FAP n will not be included in our consideration.

FU n. As well, we denote the transmit power set of FAP n on $\mathcal{S}_n$ by $\mathcal{P}_n$; that is,

$$\mathcal{P}_n = \left\{ p_n^{s_1}, p_n^{s_2}, \cdots, p_n^{s_{K_n}} \right\}, \sum_{k=1}^{K_n} p_n^{s_k} \leq p_{n,\max} \tag{5.2}$$

where $p_n^{s_k}$ represents the transmit power of FAP n on subchannel s_k and $P_{n,\max}$ is the power limit. To facilitate analysis, we do not consider power control at first and assume that $\mathcal{P}_n$ ($\forall n \in \{1, 2, \cdots, N\}$) is arbitrary, reasonable, and fixed during the underlying operational period. Meanwhile, it should be noted that both the uniform distribution mechanism (i.e., $p_n^{s_k} = \frac{P_{n,\max}}{K_n}, \forall k \in \{1, 2, \cdots, K_n\}$ and $\forall n \in \{1, 2, \cdots, N\}$) used in [41] and the simple power management schemes proposed in [28, 42] can be adopted to determine the set $\mathcal{P}_n$.

We model the additive noise as a zero-mean Gaussian random variable. For FAP n with subchannel $k \in \mathcal{S}_n$, the signal-to-interference-plus-noise ratio (SINR) can be expressed as

$$\gamma_n^k = \frac{p_n^k g_{n,n}^k}{I_n^k + B_0 N_0} = \frac{p_n^k g_{n,n}^k}{\sum\limits_{m=1, m \neq n}^{N} \sum\limits_{l \in \mathcal{S}_m \cap \{k\}} p_m^l g_{m,n}^l + B_0 N_0} \tag{5.3}$$

where I_n^k represents the cotier interference caused to FAP n on subchannel k, B_0 is the bandwidth of each subchannel, and N_0 is the noise power density. We use Shannon capacity to model the maximal achievable rate. Then, the capacity of femtocell n can be expressed as

$$R_n = \sum_{k \in \mathcal{S}_n} R_n^k = \sum_{k \in \mathcal{S}_n} B_0 \log_2 \left(1 + \gamma_n^k \right). \tag{5.4}$$

In order to maximize the capacity of the femtocell, each FAP has to choose the subchannels with lower interference. Specifically, according to Eq. (5.3) and (5.4), it can be seen that FAP n should make its decision based on other FAPs' actions and in return, the choice of FAP n will also affect other FAPs' decisions. To study the strategic interaction among these autonomous FAPs, game theory is considered as a potentially effective tool. In particular, we can formulate this subchannel selection problem as NRMG, which is defined as follows.

Definition 1. *Noncooperative rate maximization game (NRMG):*

NRMG can be represented by the tuple $\mathcal{G} = \Gamma \left(\mathcal{N}, (\mathbb{S}_n)_{n \in \mathcal{N}}, (U_n)_{n \in \mathcal{N}} \right)$, where $\mathcal{N} = \{1, 2, \cdots, N\}$ is the set of players corresponding to N FAPs. For each player n, its strategy space $\mathbb{S}_n$ is the available sets of subchannels, which can be expressed as

$$\mathbb{S}_n = \{ \mathcal{S}_n \,|\, \mathcal{S}_n \subset \{1, 2, \cdots, K\}, |\mathcal{S}_n| = K_n \} \tag{5.5}$$

where $|\cdot|$ denotes the cardinality of a set. Given a strategy profile $(\mathcal{S}_n)_{n \in \mathcal{N}} = (\mathcal{S}_1, \mathcal{S}_2, \cdots, \mathcal{S}_N) \in (\mathbb{S}_n)_{n \in \mathcal{N}}$, the utility function of each player n is

$$U_n \left((\mathcal{S}_n)_{n \in \mathbf{N}} \right) = \sum_{k \in \mathcal{S}_n} B_0 \log_2 \left(1 + \frac{p_n^k g_{n,n}^k}{I_n^k (\mathcal{S}_{-n}) + B_0 N_0} \right), \forall n \in \mathcal{N} \tag{5.6}$$

where $\mathcal{S}_{-n} = (\mathcal{S}_1, \cdots, \mathcal{S}_{n-1}, \mathcal{S}_{n+1}, \cdots, \mathcal{S}_N)$ is the strategy profile of all players other than player n and moreover, $I_n^k(\mathcal{S}_{-n})$ represents interference caused to player n by other players. Additionally, we note that the size of the strategy space of each player n is $|\mathbb{S}_n| = C_K^{K_n} = \frac{K!}{K_n!(K-K_n)!}$, where $K!$, $K_n!$, and $(K - K_n)!$ denote the factorials of K, K_n, and $K - K_n$, respectively.

According to Definition 1, the terms FAP and player will be used interchangeably hereafter. Actually, because of the conflicts among players and the absence of central authority, the social efficiency of a multiagent system will always be reduced, which is termed as PoA. We aim at devising an efficient distributed scheme to improve the social welfare (i.e., the sum capacity of femtocells $R = \sum_{n=1}^{N} R_n$). To this end, it is essential to design local learning models, with which players can update their strategies according to the environment. As shown in [43], there are some distributed learning models having been derived based on NE, which is a standard solution standing for the equilibrium state of a noncooperative game. In this light, we will next study the NE for NRMG before designing the distributed subchannel allocation strategy.

Definition 2. *Nash equilibrium (NE):*
For NRMG $\mathcal{G} = \Gamma\left(\mathcal{N}, (\mathbb{S}_n)_{n \in \mathcal{N}}, (U_n)_{n \in \mathcal{N}}\right)$, if a profile

$$\mathcal{S}^* = (\mathcal{S}_1^*, \mathcal{S}_2^*, \cdots, \mathcal{S}_N^*) \tag{5.7}$$

in the strategy space $(\mathbb{S}_n)_{n \in \mathcal{N}}$ is a NE, no player can unilaterally improve its own utility by choosing a different strategy. This means that

$$U_n\left(\mathcal{S}_n^*, \mathcal{S}_{-n}^*\right) \geq U_n\left(\mathcal{S}_n, \mathcal{S}_{-n}^*\right), \forall \mathcal{S}_n \in \mathbb{S}_n, \forall n \in \mathcal{N} \tag{5.8}$$

where $\mathcal{S}_{-n}^* = \left(\mathcal{S}_1^*, \cdots, \mathcal{S}_{n-1}^*, \mathcal{S}_{n+1}^*, \cdots, \mathcal{S}_N^*\right)$.

Unfortunately, the existence of the NE for NRMG cannot be guaranteed. To show this, here we consider a toy two-player case where $K = 2$, $N = 2$, and $K_1 = K_2 = 1$. Then, for each player $n \in \{1, 2\}$, we have $\mathcal{S}_n = \{s_n\}$, $s_n \in \{1, 2\}$, and

$$\arg\max_{s_n} U_n(s_1, s_2) = \arg\max_{s_n} \gamma_n^{s_n}(s_1, s_2). \tag{5.9}$$

In this case, the formulated NRMG is identical to the SINR maximization game introduced in [27]. Based on the numerical example presented in Table I of [27][6], we note that the SINR maximization game may have no NE. Hence, there is also no guarantee that NRMG always admits a NE. On top of this conclusion, it brings us a great challenge to design an efficient distributed subchannel allocation scheme for the formulated problem.

---

[6]Note, there is a minor notation difference between NRMG and the one formulated in [27]. Particulary, $g_{n,m}$ denotes the channel power gain from FAP n to FU m in this chapter, and $h_{n,m}^2$ denotes the channel power gain from BS m to user n in [27].

5.2.2 Distributed Algorithm Design

In this section, a quick overview of utility-based learning models is given and then the algorithm UDSA is developed. After that, the performance of the proposed scheme is mathematically analyzed.

5.2.2.1 Utility-Based Learning Models

For NRMG $\Gamma\left(\mathcal{N}, (\mathbb{S}_n)_{n\in\mathcal{N}}, (U_n)_{n\in\mathcal{N}}\right)$, at each decision time $t \in \{0, 1, 2, \cdots\}$, each player n would choose a strategy (or an action) $\mathcal{S}_n(t) \in \mathbb{S}_n$ and receive the corresponding utility $U_n\left((\mathcal{S}_n(t))_{n\in\mathcal{N}}\right)$. For individual players, their strategies will be determined by the observations from times $\{0, 1, \cdots, t-1\}$ and the pre-defined decision rule, which is referred to as the learning model. In other words, different learning models are specified by both the assumptions on available information and the rules for choosing strategies. For instance, in a best-response dynamic, given the strategies of other players, each player responds by choosing the strategy that maximizes its own utility.

In fact, there are learning models that are said to be utility-based or payoff-based [30]. In these models, it is assumed that each player can only access the history of its own actions and utilities, and players have to make decisions based on limited information. For this reason, such models are considered to be more applicable for studying multiagent systems, where the information exchange between different agents is strictly restrained. However, due to limited available information, both the convergence and efficiency of this model are often unpredictable. Recently, Li and Marden began to address this issue in their studies [31–33], and they demonstrated that it is critical to introduce an additional degree of freedom to the formulated game or to the proposed learning algorithm.

5.2.2.2 Utility-Based Distributed Subchannel Allocation

By introducing a state to reflect player's desire for new strategies, we will devise a utility-based learning model in this section. Then, our algorithm UDSA will be developed based on the proposed learning model. The detail is given as follows.

Here, to devise the utility-based learning model, we reconstruct NRMG by introducing an additional state to each player and term it as personality. In particular, players can be divided into two groups based on their personality: conservative and radical. Let us denote the personality of conservatives and radicals by c and r, respectively. According to the player's desire for new strategies, the difference between these two groups of players will be presented later. Then, at each decision moment t, every player n can be described by a triplet $\mathbb{L}_n(t) = (\mathcal{S}_n(t), U_n(t), \alpha_n(t))$, where $\mathcal{S}_n(t)$, $U_n(t)$, and $\alpha_n(t) \in \{c, r\}$ represent its strategy, utility, and personality, respectively. When implementing the utility-based learning model, during the decision period, each player only needs to evaluate its utility as well as capture the historical state and then, each player chooses a strategy. Note that each player does not need to know the strategies adopted by the opponents and in fact, it is not even aware of other players.

Considering its previous personality $\alpha_n(t-1)$ and action $\mathcal{S}_n(t-1)$ at time t, player n will first determine its mixed strategy as

$$\mathbf{Q}_n = \left(q_1, q_2, \cdots, q_{|\mathbb{S}_n|}\right) \in \Delta\left(\mathbb{S}_n\right). \tag{5.10}$$

In (5.10), for each decision epoch t, $\Delta\left(\mathbb{S}_n\right)$ denotes the set of probability distribution over the strategy space $\mathbb{S}_n$ and q_i is the probability of choosing the strategy whose index in $\mathbb{S}_n$ is i; that is

$$q_i \geq 0, \forall i \in \{1, 2, \cdots, |\mathbb{S}_n|\}, \sum_{i=0}^{|\mathbb{S}_n|} q_i = 1.$$

Hence, the mixed strategy is applied to depict the dynamics of each player during the learning process.

In particular, detailed formulas for calculating the mixed strategy are given as follows. If $\alpha_n(t-1) = r$,

$$\mathbf{Q}_n\left(i(\mathcal{F}_n)\right) = \frac{1}{|\mathbb{S}_n|}, \forall \mathcal{F}_n \in \mathbb{S}_n \tag{5.11}$$

where $i(\mathcal{F}_n)$ is the index of strategy $\mathcal{F}_n$ in $\mathbb{S}_n$, and $\mathbf{Q}_n\left(i(\mathcal{F}_n)\right)$ represents the $i(\mathcal{F}_n)$th entry in the vector $\mathbf{Q}_n$. If $\alpha_n(t-1) = c$,

$$\mathbf{Q}_n(i(\mathcal{F}_n)) = \begin{cases} \frac{\varepsilon^w}{|\mathbb{S}_n|-1}, & \forall \mathcal{F}_n \in \mathbb{S}_n, \mathcal{F}_n \neq \mathcal{S}_n(t-1) \\ 1 - \varepsilon^w & \text{otherwise} \end{cases} \tag{5.12}$$

where ε is a constant belonging to $[0, 1]$ and w is a constant larger than N. From (5.11) and (5.12) we note that the dynamics of the conservative player and those of the radical player are different. To be specific, the conservative player will occasionally adopt new strategies with a small probability, but new strategies will be frequently employed by the radical player.

After that, player n will choose an action $\mathcal{S}_n(t)$ according to the mixed strategy $\mathbf{Q}_n$, calculate its utility $U_n(t)$ by measuring interference, and then update the personality with Algorithm 5.1.

In Algorithm 1, F_n and β are constants adopted to normalize the utility; that is, $\widehat{U}_n = \left(\frac{U_n(t)}{F_n}\right)^\beta \in (0, 1), \forall t$. Here, we will show how to choose the factor F_n and defer the study of the effect of β to the next section. Actually, each FAP n can obtain the maximum achievable rate on each subchannel when there is no cotier interference; that is,

$$R_{n,\max}^k = B\log_2\left(1 + \frac{p_n^k g_{n,n}^k}{B_0 N_0}\right), \forall k \in \{1, 2, \cdots, K\}. \tag{5.13}$$

Algorithm 5.1 Personality updating algorithm

1: **if** $\alpha_n\left(t-1\right) = c$ **then**
2:　　**if** $\left(\mathcal{S}_n\left(t\right) = \mathcal{S}_n\left(t-1\right)\right)$ and $\left(U_n\left(t\right) = U_n\left(t-1\right)\right)$ **then**
3:　　　　**Set** $\alpha_n\left(t\right)$ to c
4:　　**else**
5:　　　　**Go** to Line 10.
6:　　**end if**
7: **else**
8:　　**Go** to Line 10.
9: **end if**
10: **Set** $\alpha_n\left(t\right)$ to c and r with the probability $\rho_c = \varepsilon^{1-\left(\frac{U_n(t)}{F_n}\right)^{\beta}}$ and $\rho_r = 1 - \rho_c$, respectively.

Therefore, F_n can be set as

$$F_n = \max\left\{\sum_{l\in\mathcal{S}_n} R^l_{n,\max} \,\middle|\, \mathcal{S}_n \subseteq \mathbb{S}_n\right\} \qquad (5.14)$$

which is always larger than $U_n\left(t\right), \forall t$.

Making use of the described learning model, we can develop a distributed subchannel allocation algorithm, which is termed as UDSA and depicted in Algorithm 5.2. In this algorithm, players can sequentially update their strategies. Similar to [44], the stop criterion of this algorithm can be one of the following: (1) the preset maximum iteration number T is reached or (2) for each player n, the variation of its utility during a period is trivial.

At the beginning of UDSA, the related parameters and players' states should be initialized, where $\left(\mathbf{0}\right)_{1\times M}$ represents an M-dimension null vector. After that, the algorithm goes into a loop. At each iteration t, player n will first update its state profile $\mathbb{L}_n\left(t\right) = \left(\mathcal{S}_n\left(t\right), U_n\left(t\right), \alpha_n\left(t\right)\right)$ with the devised utility-based learning model. Then, it will update the strategy count $\mathbf{C}_n$ according to its current personality. If $\alpha_n\left(t\right) = c$,

$$\mathbf{C}_n\left(i(\mathcal{S}_n\left(t\right))\right) = \mathbf{C}_n\left(i(\mathcal{S}_n\left(t\right))\right) + 1 \qquad (5.15)$$

where $i(\mathcal{S}_n\left(t\right))$ is the index of $\mathcal{S}_n\left(t\right)$ in $\mathbb{S}_n$. Intuitively, the above rule means that each player will record the strategy which makes its personality c. When the loop is exited, individual players will make their final decisions as follows

$$\mathcal{S}_n^D = \arg\max_{\mathcal{S}_n} \mathbf{C}_n\left(i(\mathcal{S}_n)\right), \forall \mathcal{S}_n \in \mathbb{S}_n, \forall n \in \mathcal{N}. \qquad (5.16)$$

From (5.16), we find that the strategy recorded most frequently will be eventually adopted.

It is seen that our proposed Algorithm 5.2 is simple and completely distributed. In particular, when each player updates its strategy, it does not require any prior information of other players, (e.g., the CSI between different

---

Algorithm 5.2 UDSA.

1: **Set** iteration count $t = 0$, personality $\alpha_n(t) = r$ and strategy count $\mathbf{C}_n = (\mathbf{0})_{1 \times |\mathbb{S}_n|}$, $\forall n \in \mathcal{N}$. Each player n randomly chooses a strategy $\mathcal{S}_n(t)$, and receives a utility $U_n(t)$.

2: **repeat**

3:     **Set** $t = t + 1$

4:     **for** $n = 1$ to N users **do**

5:         **update** state profile $\mathbb{L}_n(t)$:

6:         **if** $\alpha_n(t-1) = r$ **then**

7:             Calculate $\mathbf{Q}_n$ with Eq. (5.11).

8:         **else**

9:             Calculate $\mathbf{Q}_n$ with Eq. (5.12).

10:         **end if**

11:         **Choose** a strategy $\mathcal{S}_n(t)$, measure the utility $U_n(t)$, and update its personality $\alpha_n(t)$.

12:         **Update** strategies count $\mathbf{C}_n$:

13:         **if** $\alpha_n(t) = c$ **then**

14:             Update $\mathbf{C}_n$ with Eq. (5.15).

15:         **end if**

16:     **end for**

17: **until** the stop criterion is satisfied.

18: Each player decides its strategy according to (5.16).

---

femtocells and the utility functions of its competitors). Next, we will analyze the complexity of this algorithm.

Recalling the proposed algorithm UDSA that can be implemented in parallel, we note that each player only needs to make its own decision and meanwhile, only basic arithmetic operations and random number generation are involved in each iteration step. Therefore, the complexity of this algorithm is dependent on both the stop criterion of the loop and the sizes of players' strategy spaces. In particular, for the two different stop criterions earlier described, the complexities are $O(T + L)$ and $O(E + L)$, respectively, where T is the preset maximum iteration number, $L = \max\{|\mathbb{S}_1|, |\mathbb{S}_2|, \cdots, |\mathbb{S}_N|\}$, and E is the convergence rate of the algorithm. In addition, it should be noted that the convergence rate E is related to the algorithm parameters, which will be further discussed in the following section.

5.2.2.3 Performance Analysis of UDSA

In this section, the performance of the developed algorithm will be analyzed. For NRMG, let $\left(\mathcal{S}_n^O\right)_{n \in \mathcal{N}} \in (\mathbb{S}_n)_{n \in \mathcal{N}}$ denote the profile of strategies satisfying

$$\left(\mathcal{S}_n^O\right)_{n \in \mathcal{N}} = \underset{(\mathcal{S}_n)_{n \in \mathcal{N}}}{\arg\max} \, \hat{U}\left((\mathcal{S}_n)_{n \in \mathcal{N}}\right) \tag{5.17}$$

where

$$\hat{U}\left((\mathcal{S}_n)_{n\in\mathcal{N}}\right) = \sum_{n=1}^{N} \hat{U}_n\left((\mathcal{S}_n)_{n\in\mathcal{N}}\right) = \sum_{n=1}^{N} \left(\frac{U_n\left((\mathcal{S}_n)_{n\in\mathcal{N}}\right)}{F_n}\right)^{\beta}. \tag{5.18}$$

According to [45], we can show that UDSA can asymptotically converge to the solution maximizing the aggregate normalized utility under the given condition. In particular, When $\left(\mathcal{S}_n^O\right)_{n\in\mathcal{N}}$ is unique and ε is sufficiently small ($\varepsilon \to 0$), UDSA asymptotically converges to $\left(\mathcal{S}_n^O\right)_{n\in\mathcal{N}}$, i.e.

$$\mathrm{Pr}\left(\lim_{T\to\infty,\varepsilon\to 0}\left(\mathcal{S}_n^D\right)_{n\in\mathbf{N}} = \left(\mathcal{S}_n^O\right)_{n\in\mathcal{N}}\right) = 1 \tag{5.19}$$

where T is the number of iterations.

In general, for distributed algorithms, the effectiveness of the achieved solutions are generally evaluated using Pareto optimality or known as Pareto efficiency, which is introduced in the following definition.

Definition 3. *Pareto optimality (Pareto efficiency):*

For a situation, a profile

$$\mathcal{S}^{PO} = \left(\mathcal{S}_1^{PO}, \mathcal{S}_2^{PO}, \cdots, \mathcal{S}_N^{PO}\right)$$

in the strategy space is Pareto-optimal (Pareto-efficient), if and only if there is no other set of strategies for which at least one player can improve its own welfare without reducing those of other users.

We refer to the condition that $\varepsilon \to 0$ and $\left(\mathcal{S}_n^O\right)_{n\in\mathbf{N}}$ is unique as the ideal condition. Then, for NRMG $\Gamma\left(\mathcal{N}, (\mathbb{S}_n)_{n\in\mathcal{N}}, (U_n)_{n\in\mathcal{N}}\right)$, the solution achieved by UDSA, $\left(\mathcal{S}_n^D\right)_{n\in\mathcal{N}}$, is Pareto-optimal when the ideal condition is met. This is true accoding to reductio ad absurdum. We assume that $\left(\mathcal{S}_n^D\right)_{n\in\mathcal{N}}$ is not Pareto-optimal. Mathematically speaking, $\exists\left(\tilde{\mathcal{S}}_n\right)_{n\in\mathcal{N}} \in \mathbb{S}_n :$

$$\hat{U}_n\left(\left(\tilde{\mathcal{S}}_n\right)_{n\in\mathcal{N}}\right) \geq \hat{U}_n\left(\left(\mathcal{S}_n^D\right)_{n\in\mathcal{N}}\right), \forall n \in \mathcal{N} \tag{5.20}$$

and meanwhile,

$$\hat{U}_m\left(\left(\tilde{\mathcal{S}}_n\right)_{n\in\mathcal{N}}\right) > \hat{U}_m\left(\left(\mathcal{S}_n^D\right)_{n\in\mathcal{N}}\right), \exists m \in \mathcal{N}. \tag{5.21}$$

Then, we will get

$$\sum_{n=1}^{N} \left(\frac{U_n\left(\left(\tilde{\mathcal{S}}_n\right)_{n\in\mathcal{N}}\right)}{F_n}\right)^{\beta} > \sum_{n=1}^{N} \left(\frac{U_n\left(\left(\mathcal{S}_n^D\right)_{n\in\mathcal{N}}\right)}{F_n}\right)^{\beta} \tag{5.22}$$

which contradicts with (5.19). Therefore, $\left(\mathcal{S}_n^D\right)_{n\in\mathcal{N}}$ is Pareto-optimal.

Accordingly, when $\left(\mathcal{S}_n^{PO}\right)_{n\in\mathcal{N}}$ is achieved, the overall capacity

$$R = \sum_{n=1}^{N} R_n = \sum_{n=1}^{N} U_n \tag{5.23}$$

cannot be further improved without reducing the capacity of any one femtocell. In other words, this solution will result in a win-win outcomes for both the femtocell owners and operator. We also note that there is no requirement that $\left(\mathcal{S}_n^{PO}\right)_{n\in\mathcal{N}}$ is the NE for NRMG. Otherwise stated, it can happen that this efficient point may be ignored by the distributed scheme whose aim is to reach a NE for the game.

As the end of this section, we shift our focus to study the effects of the parameters β and ε that are adopted in the proposed algorithm. Actually, when the parameter profile $\{F_1, F_2, \cdots, F_N\}$ is selected, the concerned Pareto-optimal solution $\left(\mathcal{S}_n^{PO}\right)_{n\in\mathcal{N}}$ can also be achieved by solving the following integer programming:

$$\underset{(\mathcal{S}_n)_{n\in\mathcal{N}}}{\arg\max} \quad \widehat{U} = \sum_{n=1}^{N} \widehat{U}_n \tag{5.24}$$

$$= \sum_{n=1}^{N} \left(\frac{R_n\left((\mathcal{S}_n)_{n\in\mathcal{N}}\right)}{F_n} \right)^{\beta},$$

$$s.t. : \quad \mathcal{S}_n \subset \{1, 2, \cdots, K\},$$

$$|\mathcal{S}_n| = K_n, \forall n \in \{1, 2, \cdots, N\}.$$

For notational simplicity, we denote by $\widehat{U}_{PO}$ and R_{PO} the corresponding aggregate normalized utility and capacity, respectively; that is,

$$\widehat{U}_{PO} = \sum_{n=1}^{N} \widehat{U}_n \left(\left(\mathcal{S}_n^{PO}\right)_{n\in\mathcal{N}}\right) \tag{5.25}$$

and

$$R_{PO} = \sum_{n=1}^{N} R_n \left(\left(\mathcal{S}_n^{PO}\right)_{n\in\mathcal{N}}\right). \tag{5.26}$$

It can be seen from the problem shown in (5.24) that the value of β will characterize both the utility $\widehat{U}_{PO}$ and the capacity R_{PO}. Specifically, $\widehat{U}_{PO}$ is monotonically decreasing with respect to β. Nevertheless, the relationship between β and R_{PO} is not obvious and intractable for mathematical analysis.

On the other hand, both the values of β and ε will affect the convergence of UDSA. Smaller ε will lead to a slower convergence speed, but the algorithm is more likely to converge to the Pareto-optimal solution $\left(\mathcal{S}_n^{PO}\right)_{n\in\mathcal{N}}$. In addition,

Table 5.1: Simulation Parameters

Parameter	Value
Number of subchannels, K	5
Region radius, r	100 m
Maximum distance between each FAP-FU pair, D	10 m
Bandwidth per subchannel, B_0	100 kHz
Subchannel requirement, K_n, $\forall n$	3
FAP power limit, $p_{n,max}$, $\forall n$	20 dBm
Standard deviation of ψ, σ_ψ	4 dB
Wall penetration loss, L_{wall}	5 dB
Noise figure, φ_{FU}	7 dB
AWGN power density, N_0	-174 dBm/Hz

the value of β will also affect the performance of UDSA. As shown in line 10 of Algorithm 5.1, smaller β will make players set their personality to c more frequently. This updating rule eventually implies that each player n may act irrationally and that the algorithm will converge faster. The above comments will be illustrated with simulation results.

5.2.3 Results and Analysis

5.2.3.1 Simulation Scenario

We consider a circular region of radius r m, where N femtocells are randomly deployed. In each femtocell, the distance between a FAP-FU pair is a uniform random variable between 0 and D m. Furthermore, for each FAP-FU pair, the channel gains are independent on different subchannels. During the transmission period, we employed the uniform distribution mechanism to determine the power set; that is,

$$p_n^l = \frac{p_{n,\max}}{K_n}, \forall l \in \mathcal{S}_n, \forall \mathcal{S}_n \in \mathbb{S}_n, \forall n \in \mathcal{N}.$$

Unless specified otherwise, the simulation parameters are shown in Table 5.1[7] [28,40] and each individual simulation result is obtained by averaging over 10,000 independent runs.

---

[7] We considered internal building walls and hence the building penetration loss is relatively small.

Table 5.2: The Value of $\widehat{U}_{PO}$ and R_{PO} versus β

β	$\widehat{U}_{PO}$	R_{PO} (Mbps)	β	$\widehat{U}_{PO}$	R_{PO} (Mbps)
0.1	6.64	15.76	0.2	6.40	15.93
0.3	6.07	15.96	0.4	5.88	16.02
0.5	5.60	15.46	0.6	5.32	15.83
0.7	5.30	15.79	0.8	5.03	15.88
0.9	4.93	15.54	1.0	4.90	16.01

5.2.3.2   Convergence

To verify the validity of our analysis and evaluate the convergence of the proposed algorithm, we first calculate the values of $\widehat{U}_{PO}$ and R_{PO} when β is set to different values. Then, we will explore the difference between $\widehat{U}_{PO}$ and $\widehat{U}_{D}$, which is the normalized utility achieved by UDSA. Note that there is no efficient algorithm to solve the integer programming shown in (5.24), and hence the method of exhaustion is used in this section, which needs to compare all the $\prod_{n\in\mathcal{N}} |\mathbb{S}_n|$ utilities to obtain the optimal solution. To find $\left(\mathcal{S}_n^{PO}\right)_{n\in\mathcal{N}}$ within an acceptable period of time, we consider a small-scale scenario as an example, where 7 femtocells are deployed in a random fashion and for each FU n $K_n = 1$.

Table 5.2 demonstrates the value of $\widehat{U}_{PO}$ and R_{PO} as β changes from 0.1 to 1. It is obvious that the larger β leads to the smaller $\widehat{U}_{PO}$. However, the variation pattern of R_{PO} is difficult to characterize, which is consistent with our previous comments in Section 5.2.2.3. For instance, $R_{PO}\,(\beta = 0.1) < R_{PO}\,(\beta = 0.3)$, $R_{PO}\,(\beta = 0.3) > R_{PO}\,(\beta = 0.5)$, but $R_{PO}\,(\beta = 0.5) < R_{PO}\,(\beta = 0.7)$. Then, we illustrate the convergence of UDSA with Figure 5.1, where β is set to 0.1 and 0.2, respectively.

It is seen that when β and ε are given, the development of our algorithm follows a monotonically increasing path before stabilizing. As well, smaller ε results in longer convergence time for given β, but the stable $\widehat{U}_{D}$ is closer to $\widehat{U}_{PO}$, which is illustrated by the solid curve. For example, in Figure 5.1 (a), when $\varepsilon = 10^{-1}$ and $\varepsilon = 10^{-5}$ the convergence time is about 10 and 100 iterations, respectively. In addition, when ε is set to 10^{-5}, the achieved $\widehat{U}_{D}$ is approximately 3.95% higher than that of $\varepsilon = 10^{-1}$. Meanwhile, we note that in the case that $\beta = 0.1$ and $\varepsilon = 10^{-5}$, there still is a small difference between $\widehat{U}_{D}$ and $\widehat{U}_{PO}$, and specifically, the relative gap is around 0.81%. Theoretically, this difference is caused by two reasons: first, ε is not small enough, and second, there is no guarantee that the Pareto-optimal solution is always unique in each round of simulation. We would also note that the convergence speed is reduced with the increase of β when ε is given. For instance, when $\varepsilon = 10^{-5}$ and $\beta = 0.1$, the convergence rate is nearly 10 times faster than that in the case where $\varepsilon = 10^{-5}$ and $\beta = 0.2$.

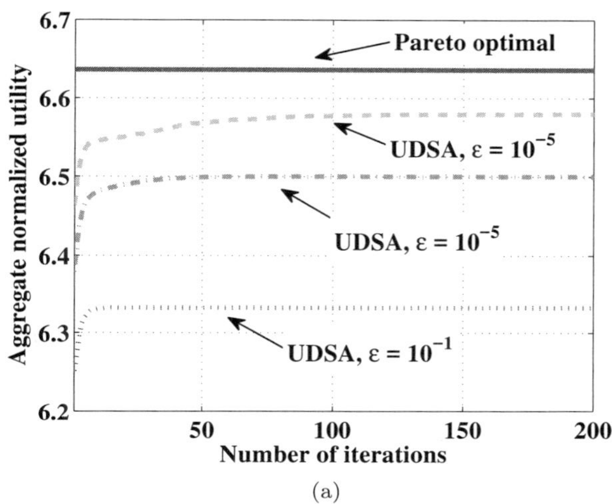

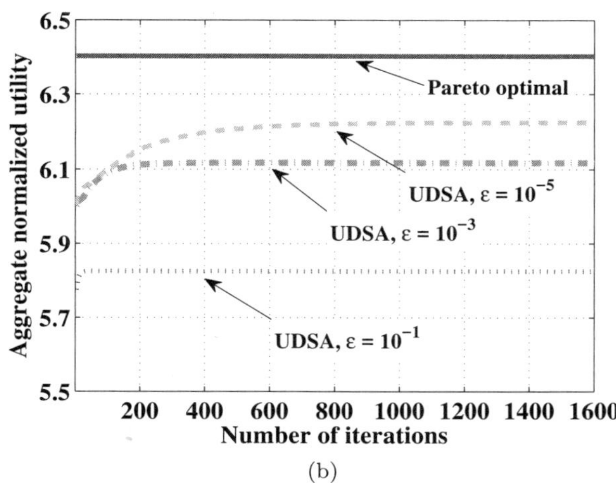

Figure 5.1: The convergence of UDSA with respect to ε for (a) $\beta = 0.1$ and (b) $\beta = 0.2$.

Table 5.3: Information Requirement of DSA Schemes

Schemes	DSA-[27]	DSA-[28]	UDSA
CSI between different cells	✓	✓	✗
Other players' strategies	✓	✓	✗
Other players' transmit power	✓	✓	✗
Incoming interference	✗	✗	✓

5.2.3.3 Performance Evaluation

To demonstrate interference mitigation capability of UDSA, two distributed subchannel allocation (DSA) schemes are compared, which have been proposed in Section IV of [27] and [28], respectively. For notional simplicity, we denote them as DSA-[27] and DSA-[28]. On one hand, we compare them in terms of the required information, which is shown in Table 5.3. In particular, using DSA-[27] and DSA-[28], each FAP has to know the CSI from other FAPs to its own FU and that from itself to the FUs associating to other FAPs. In addition, for each individual FAP, it is necessary to know the strategies adopted by the other FAPs to choose its own best response action in the each iteration of the developed algorithm. However, one FAP only needs to measure the incoming interference and calculate its own utility when implementing UDSA. As a result, no information exchange is required among different femtocells.

On the other hand, we investigate the performance of UDSA in terms of the average SINR per subchannel in decibels

$$\bar{\gamma} = \frac{1}{N} \sum_{n=1}^{N} \frac{\sum_{l \in \mathcal{S}_n} 10\log_{10}\left(\gamma_n^l\right)}{K_n} \tag{5.27}$$

and of the overall capacity R, which are illustrated in Figure 5.2(a) and 5.2(b), respectively. In this simulation, femtocells are randomly deployed and moreover, K_n, β, and ε are set to 3, 0.1, and 0.01, respectively. As well DSA-[27] and DSA-[28], there is another method which should be compared, with which each FAP n will randomly utilizes subchannels. We term this strategy as DSA-random and regard its performance as the baseline.

As shown in Figure 5.2(a), the average SINR decreases when there are more active femtocells. This is because that more femtocells will result in higher interference. Furthermore, we note that although there is no information exchange among the FAPs, the proposed algorithm can bring the highest SINR. In other words, interference mitigation capability of our strategy is better than that of existing strategies. Compared with DSA-random, DSA-[27], and DSA-[28], UDSA has around 45.7%, 16.8%, and 9.8% higher average SINR, respectively.

From the perspective of overall capacity, we have compared UDSA with the available strategies in different interference environments (i.e., the number of communicating femtocells N is set to different values). As demonstrated in

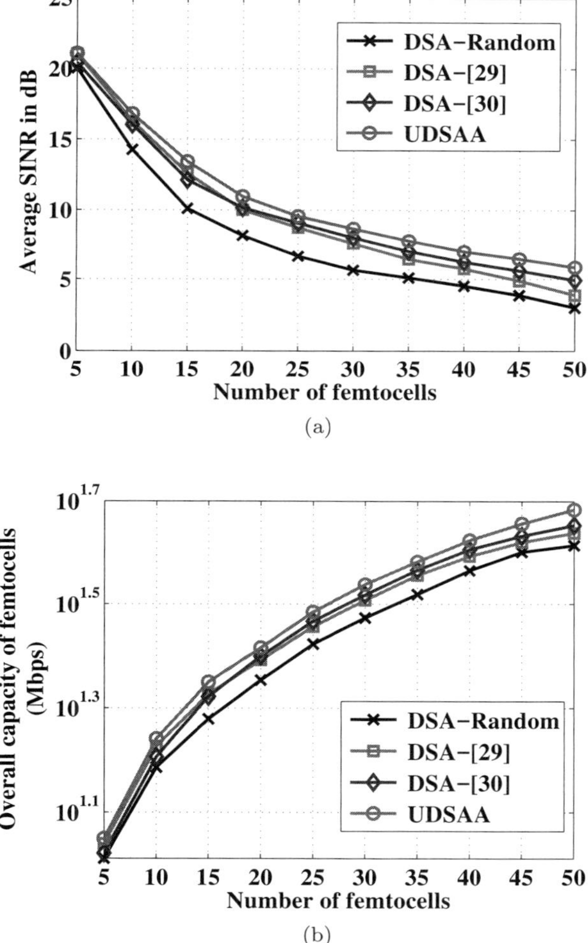

Figure 5.2: Performance comparison in terms (a) of the average SINR $\bar{\gamma}$, and (b) of the overall capacity R.

Figure 5.2(b), our scheme always brings higher sum capacity, which is accordant with the result depicted in Figure 5.2(a). For instance, when $N = 50$, UDSA provides approximately 11.0% more capacity than that of DSA-[27] and 7.2% more capacity than that of DSA-[28]. The reason for this improvement is that the purpose of each player is to maximize its capacity. However, in both DSA-[27] and DSA-[28], the relationship between the utility and capacity is not obvious. Meanwhile, from the simulation results we note that the rate of improvement of the overall capacity decreases when the density of femtocells becomes high. The similar observation has also been made in previous work [46], where it is referred to as the fundamental throughput scaling limit.

5.2.3.4   A More Realistic Scenario

In this section we demonstrate the performance of our algorithm in more realistic LTE femtocells, where the system bandwidth is 3 MHz (i.e., $K = 15$) and the bandwidth of each RB is 180 KHz. Throughout this section, for every user $n \in \mathcal{N}$ we consider $K_n = \frac{K}{5} = 3$ (i.e., the size of strategy space $|\mathbb{S}_n|$ is 455). In addition, the simulation result is obtained by averaging over 1,000 independent runs.

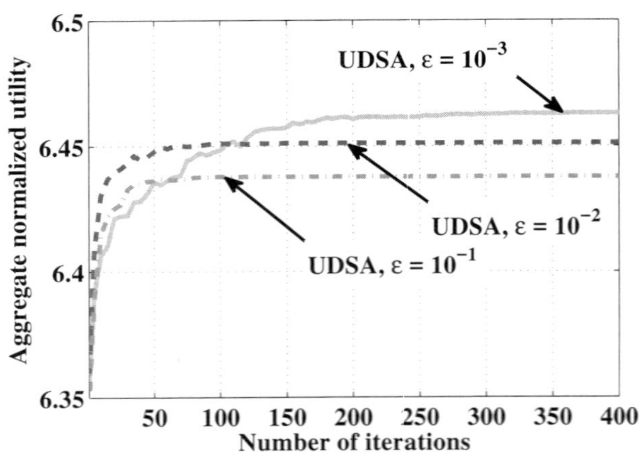

Figure 5.3: The convergence of UDSA with respect to ε for $\beta = 0.1$.

To compare with the results shown in Figure 5.1, here we also set $N = 7$ and $\beta = 0.1$ and then illustrate the convergence of our algorithm in Figure 5.3. We note that, in contrast with Figure 5.1, the Pareto-optimal utility $\widehat{U}_{PO}$ is not shown here. The reason is that $\widehat{U}_{PO}$ can only be obtained after comparing the all $\prod_{n \in \mathcal{N}} |\mathbb{S}_n| = 4.037 \times 10^{18}$ strategy profiles, but such a huge amount of enumeration cannot be performed in reasonable time. From the simulation results it is seen that a trade-off between efficiency and convergence rate can also be made by adjusting the parameter ε. In addition, compared with the results presented in Figure 5.1, we note that a larger size of strategy for each

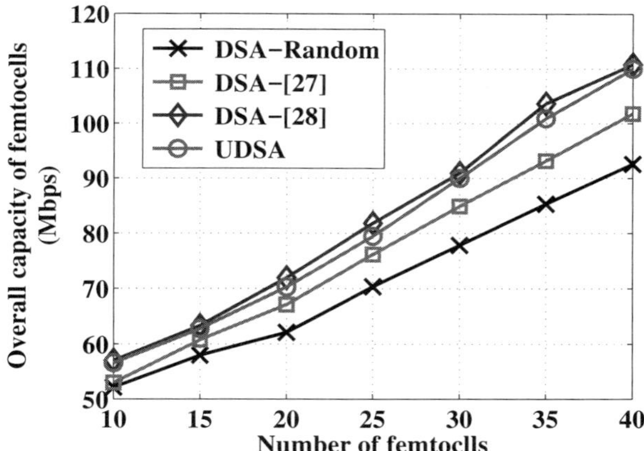

Figure 5.4: Performance comparison in term of overall capacity R for LTE femtocells, where the total bandwidth is 3 MHz.

player will make the convergence of our algorithm slower. This is due to the fact that the subchannel allocation is essentially a combinatorial problem. A similar observation can also be made in previous distributed algorithms [27, 28, 44].

Finally, we compare the performance of our scheme with the three available ones (i.e., DSA-Random, DSA-[27], and DSA-[28]). In particular, as the number of femtocells N increases, the performance in terms of overall capacity R is demonstrated in Figure 5.4, where β and ε are set to 0.1 and 0.01, respectively. From the simulation results we can see that our devised method has roughly the same performance as DSA-[28] but without information exchange among different femtocells. On the other hand, compared to DSA-[27], even better performance can be obtained by our scheme. For instance, when $N = 40$ UDSA yields a performance advantage around 8.2% relative to the scheme DSA-[27] (i.e., from 101.62 Mbps to 110.04 Mbps). It should be noted that the reasons for this improvement are similar to those described previously and moreover, the simulation results presented here also show the advantage of our approach.

5.3   Distributed Power Control and Subchannel Allocation

5.3.1   System Model and Problem Formulation

We consider a similar scenario as in the previous section, where there are in total N femtocells sharing the dedicated spectrum consisting of K orthogonal subchannels. The bandwidth of each subchannel is B_0. We consider that each channel can be assigned to multiple femtocells and meanwhile, interference occurs when each channel is simultaneously utilized by more than one femtocell. Without loss of generality, we suppose $N \geq K$. For notational simplicity, let vectors $\mathcal{N}$ and $\mathcal{K}$ denote the set of FAPs and channels, respectively (i.e., $\mathcal{N} = \{1, 2, \cdots, N\}$ and $\mathcal{K} = \{1, 2, \cdots, K\}$). Additionally, we denote the channel selected by FAP n by $c_n \in \mathcal{K}$. Also, no CCC or central authority is assumed for coordination among FAPs.

Let $\mathbf{G} \in \mathbb{R}^{N \times N \times K}$ be the channel power gain matrix, where $g_{n,m}^k$ represents the channel gain between transmitter n and receiver m on channel k. We assume the channel condition is static during the underlying operational period (e.g., the quasi-static scenario). The additive noise is modelled as a zero-mean Gaussian random variable, and then, for user n, its signal-to-interference-plus-noise ratio (SINR) can be expressed as

$$\gamma_n = \frac{p_n g_{n,n}^{c_n}}{I_n^{c_n} + B_0 N_0} = \frac{p_n g_{n,n}^{c_n}}{\sum\limits_{m \in \mathcal{N}, m \neq n} \delta(c_m, c_n) p_m g_{m,n}^{c_n} + B_0 N_0}, \tag{5.28}$$

where I_n represents interference caused to user n, p_n is the transmit power of FAP n, and N_0 is the noise power density. As well that, the indicative function $\delta(c_m, c_n)$ is adopted to indicate whether the same channel is used by FAP m and n simultaneously or not: if $c_m = c_n$, $\delta(c_m, c_n) = 1$; otherwise $\delta(c_m, c_n) = 0$. In the following, we consider that each FAP n can choose the transmit power p_n from a finite set $\mathcal{P}_n = \{p_n^1, p_n^2, \cdots, p_n^{max}\}$ [35, 36].

Based on the above, the achievable transmission rate of FAP n can be expressed as

$$R_n = B_0 \log_2\left(1 + \gamma_n\right). \tag{5.29}$$

Adopting different channels and power levels, one FAP will obtain different achievable rates. According to (5.28) and (5.29), if FAP n transmits on channel c_n, R_n can be maximized with power p_n^{max} when there is no interference. Therefore, the upper bound of the rate R_n for FAP n can be defined as

$$R_n^{max} = \max\left\{ B_0 \log_2\left(1 + \frac{p_n^{max} g_{n,n}^{c_n}}{B_0 N_0}\right) \Big| c_n \in \mathcal{K} \right\}. \tag{5.30}$$

Moreover, we consider that each user n has rate requirement R_n^{min} to satisfy its QoS requirement and assume that $0 \leq R_n^{min} \leq R_n^{max}$.

Intuitively, in this network not all the users' rate requirements can be guaranteed when they transmit simultaneously, especially for the case where all the users' rate requirements are high [35]. For instance, if $R_n^{min} = R_n^{max}, \forall n \in \mathcal{N}$, then there are at most K transmissions being permitted. Here, to get around this problem, we consider softening the user's rate requirement and measure its degree of satisfaction with a sigmoid function.[8] In fact, this approach has been widely adopted in radio resource management [47–50]. To this end, the utility of each individual user can be expressed as

$$U_n(R_n) = \frac{1}{1 + e^{-\beta_n(R_n - R_n^{min})}}, \forall n \in \mathcal{N}, \tag{5.31}$$

---

[8] We defined the degree of user satisfaction as a function of user's rate requirement. Therefore, it can be used to model data traffic.

where β_n is a constant deciding the steepness of the satisfactory curve, and moreover, both R_n and R_n^{min} are considered to have units in Mbps. It is clear from the above equation that $U_n(R_n)$ is a monotonic increasing function with respect to R_n (i.e., individual users will feel more satisfied when they have higher rate). Furthermore, since $\lim_{R_n \to 0} U_n(R_n) = \frac{1}{1+e^{\beta_n R_n^{min}}} > 0$ and $\lim_{R_n \to \infty} U_n(R_n) = 1$, the utility of each user n is scaled between 0 and 1 (i.e., $U_n(R_n) \in (0,1)$). We note that although the higher utility means the higher spectral efficiency for a given bandwidth, the value of the former cannot directly reflect the value of the latter. Therefore, in simulation results, not only the overall utility U but also the average rate $\bar{R}$ are recorded and shown to evaluate the efficiency of different algorithms.

Before starting a transmission, each individual FAP should decide to adopt which power level and transmit on which channel. For notational simplicity, we refer to a pair of channel index and power level as a strategy $\mathbf{s}_n$; that is,

$$\mathbf{s}_n = (c_n, p_n) \in \mathcal{S}_n, \mathcal{S}_n = \mathcal{K} \times \mathcal{P}_n, \forall n \in \mathcal{N}. \tag{5.32}$$

From (5.28), (5.29), and (5.31), we note that each user's rate is affected by the transmissions of other users and meanwhile, a higher rate brings higher satisfaction to a user. Therefore, to improve the degree of satisfaction or utility, each user should choose its own strategy by considering the actions of other users. That is, there is a coupling among the strategies employed by different users. In order to well study the conflict among different users, NTCG has been formulated, and hereafter, the terms user and player will be used interchangeably.

Definition 4. *NTCG:*

NTCG can be represented by the tuple

$$\mathcal{G} = \Gamma\left(\mathcal{N}, (\mathcal{S}_n)_{n \in \mathcal{N}}, (U_n)_{n \in \mathcal{N}}\right). \tag{5.33}$$

In particular, $\mathcal{N}$ denotes the set of players that is identical to the user set. For each player n, its strategy space $\mathcal{S}_n$ is defined as shown in (5.32). Given a strategy profile

$$(\mathbf{s}_n)_{n \in \mathcal{N}} = (\mathbf{s}_1, \mathbf{s}_2, \cdots, \mathbf{s}_N) \in (\mathcal{S}_n)_{n \in \mathcal{N}} \tag{5.34}$$

the utility function of each player n is

$$U_n((\mathbf{s}_n)_{n \in \mathcal{N}}) = U_n(R_n((\mathbf{s}_n)_{n \in \mathcal{N}})), \forall n \in \mathcal{N}, \tag{5.35}$$

where $R_n((\mathbf{s}_n)_{n \in \mathcal{N}})$ represents the achievable rate when player n adopts the strategy $\mathbf{s}_n = (c_n, p_n)$; that is,

$$R_n((\mathbf{s}_n)_{n \in \mathcal{N}}) = \log_2(1 + \frac{p_n g_{n,n}^{c_n}}{I_n^{c_n}(\mathbf{s}_{-n}) + B_0 N_0}). \tag{5.36}$$

In (5.36), $\mathbf{s}_{-n} = (\mathbf{s}_1, \cdots, \mathbf{s}_{n-1}, \mathbf{s}_{n+1}, \cdots, \mathbf{s}_N)$ is the strategy profile of all players other than player n, and $I_n^{c_n}(\mathbf{s}_{-n})$ represents interference caused to player n on channel c_n.

Obtaining the optimal channel selection and power control strategy for this decentralized network is equivalent to solving the following combinatorial problem $\mathbf{P}$, which is NP-hard.

$$\mathbf{P}: \max_{\mathbf{c},\mathbf{p}} \sum_{n\in\mathcal{N}} U_n\left((c_n,p_n)_{n\in N}\right) \tag{5.37}$$

$$s.t. \quad \mathbf{c}\in\{(c_1,c_2,\cdots,c_N)\,|\forall c_n\in\mathcal{K},\forall n\in\mathcal{N}\}, \tag{5.38}$$

$$\mathbf{p}\in\{(p_1,p_2,\cdots,p_N)\,|\forall p_n\in\mathcal{P}_n,\forall n\in\mathcal{N}\}. \tag{5.39}$$

The objective function (5.37) means that our objective is to maximize the social welfare or overall utility, which is determined by both the achievable and required rate of users (i.e., $(R_1,R_2,\cdots,R_N)$ and $\left(R_1^{min},R_2^{min},\cdots,R_N^{min}\right))$. In addition, constraints (5.38) and (5.39) specify each individual user's available channel and power level sets, respectively.

Unfortunately, the above problem is an integer programming which is extremely difficult to solve. Moreover, due to the fact that there is no central authority controlling the users in this decentralized network, developing a completely distributed algorithm to obtain the optimal solution of $\mathbf{P}$ is important and nontrivial.

5.3.2   Distributed Algorithm Design

In this section, we develop a utility-based algorithm for NTCG to achieve the solution of $\mathbf{P}$ shown in Section 5.3.1. We first prove the uncertainty of the existence of NE for NTCG, and then develop a utility-based algorithm. At the end of this section, we will investigate the complexity of the proposed algorithm and finally, prove that our algorithm can asymptotically converge to the global optimal solution under the given condition, regardless of whether this solution is a NE of the formulated game.

5.3.2.1   NE for NTCG

Recalling that there is no CCC for exchanging information among different players, the utility-based learning algorithm is therefore considered to be more appropriate for this distributed environment. Actually, in the recent work [35], a similar problem has also been studied and moreover, an efficient utility-based learning algorithm has been proposed. In particular, for the formulated game, authors in [35] has proved that if a NE exists that can maximize the social welfare, this NE can be achieved with their proposed distributed algorithm. In order to check whether the algorithm devised in [35] also can be adopted to solve our problem $\mathbf{P}$, we first discuss the existence of NE for NTCG.

For a noncooperative game, (pure-strategy) NE is a standard solution standing for the equilibrium state, under which no player can unilaterally improve its own utility by choosing a different strategy [51]. Mathematically speaking, if a profile $\mathbf{s}^* = (\mathbf{s}_1^*,\mathbf{s}_2^*,\cdots,\mathbf{s}_N^*)$ in the strategy space $(\mathcal{S}_n)_{n\in\mathcal{N}}$ is a NE, then we have

$$U_n\left(\mathbf{s}_n^*,\mathbf{s}_{-n}^*\right) \geq U_n\left(\mathbf{s}_n,\mathbf{s}_{-n}^*\right), \forall \mathbf{s}_n \in \mathcal{S}_n, \forall n \in \mathcal{N}, \tag{5.40}$$

where $\mathbf{s}_{-n}^* = \left(\mathbf{s}_1^*, \cdots, \mathbf{s}_{n-1}^*, \mathbf{s}_{n+1}^*, \cdots, \mathbf{s}_N^*\right)$.

Theorem 1. *There is no guarantee that the NE for NTCG always exists.*

Proof. For each player n, we have

$$\begin{aligned}
&\arg\max_{\mathbf{s}_n} U_n\left(\mathbf{s}_n, \mathbf{s}_{-n}\right) \\
&= \arg\max_{\mathbf{s}_n} \frac{1}{1+e^{-\omega_n\left(R_n\left(\mathbf{s}_n,\mathbf{s}_{-n}\right)-R_n^{min}\right)}} \\
&= \arg\max_{\left(p_n^{max},c_n\right)} R_n\left(\left(p_n^{max},c_n\right),\mathbf{s}_{-n}\right) \\
&= \arg\max_{\left(p_n^{max},c_n\right)} \gamma_n\left(\left(p_n^{max},c_n\right),\mathbf{s}_{-n}\right). \tag{5.41}
\end{aligned}$$

It is noted from the above equation that NTCG is identical to the SINR-maximization game introduced in [52]. According to the conclusion drawn from a toy two-user case in [52], the existence of the NE for the SINR-maximization game cannot be guaranteed, which further indicates that the NE for NTCG may also not exist. Note that a counterexample can be easily derived with the parameters given in table I in [52], and hence it is omitted here.

Now, the proof is complete. □

To this end, we can see that the utility-based algorithm developed in [35] cannot be directly applied to solve the problem addressed in this work. Hence, a novel utility-based algorithm will be devised in the following section.

5.3.2.2 Utility-Based Distributed Transmission Control Algorithm

When devising a utility-based learning algorithm, there are two components that should be elaborated for each player: the state profile and learning model (dynamics) [21,30]. More specifically, the former depicts each player's available local information, and the latter tells the users how to make their decisions based on this information. In this section, we first define the proper state profile and learning model for each player in NTCG in detail. Then, a utility-based distributed transmission control algorithm is proposed.[9]

State profile At each decision moment $t \in \{1, 2, \cdots\}$, we consider describing the state profile of player n with a triplet $\mathbb{L}_n\left(t\right) = \left(\mathbf{s}_n\left(t\right), U_n\left(t\right), \alpha_n\left(t\right)\right)$, where $S_n\left(t\right)$, $U_n\left(t\right)$, and $\alpha_n\left(t\right) \in \{0,1\}$ represent its strategy, utility, and mood,

---

[9]The utility-based learning approach implemented in this chapter can also be viewed as a state-based learning approach. The term "utility-based" is adopted by research found in [21,30] and the references therein.

respectively. We note that the binary variable $\alpha_n(t)$ is used to elaborate players' desire for changing the currently adopted strategy, which will be specified in detail when introducing the learning model.

Learning model Motivated by Marden's work [32], a utility-based learning model is adopted in this chapter, with which each individual player n can update its $\mathbf{s}_n(t)$, $U_n(t)$ and $\alpha_n(t)$ in sequence at each decision moment t. To be specific, at the beginning of time t, individual player n first needs to determine the probability distribution over the set of its available strategies (i.e., mixed-strategy)

$$\mathbf{Q}_n(t) = \left(q_n^1(t), q_n^2(t), \cdots, q_n^{|\mathcal{S}_n|}(t) \right), \tag{5.42}$$

where $|\cdot|$ represents the cardinality of a set, and $q_n^j(t)$ is the probability of choosing the jth strategy at time t, that is,

$$q_n^j(t) \geq 0, \forall j \in \{1, 2, \cdots, |\mathcal{S}_n|\}, \sum_{j=1}^{|\mathcal{S}_n|} q_n^j(t) = 1. \tag{5.43}$$

In other words, the probability distribution $\mathbf{Q}_n(t)$ is adopted to describe the players' dynamics. Here, player n would update $\mathbf{Q}_n(t)$ based on its previous mood $\alpha_n(t-1)$ and action $\mathbf{s}_n(t-1)$. In particular, if $\alpha_n(t-1) = 0$,

$$q_n^{i(\mathbf{f}_n)}(t) = \frac{1}{|\mathcal{S}_n|}, \forall \mathbf{f}_n \in \mathcal{S}_n, \tag{5.44}$$

where $i(\mathbf{f}_n)$ denotes the index of strategy $\mathbf{f}_n$ in $\mathcal{S}_n$. The rule shown in (5.44) means that if the previous mood is 0 the player will choose each strategy with equal probability. On the other hand, if $\alpha_n(t-1) = 1$

$$q_n^{i(\mathbf{f}_n)}(t) = \begin{cases} \frac{\varepsilon^w}{|\mathcal{S}_n|-1}, & \forall \mathbf{f}_n \in \mathcal{S}_n, \mathbf{f}_n \neq \mathbf{s}_n(t-1) \\ 1 - \varepsilon^w, & \text{otherwise} \end{cases}, \tag{5.45}$$

where ε is a constant belonging to $(0, 1)$ and w is a constant greater than N. The above equation means that if the previous mood is 1 then the player will change its strategy to a different one (i.e., $\mathbf{f}_n \neq \mathbf{s}_n(t-1)$) with probability $\frac{\varepsilon^w}{|\mathcal{S}_n|-1}$. Meanwhile, the same strategy (i.e., $\mathbf{f}_n = \mathbf{s}_n(t-1)$) will be adopted with probability $1 - \varepsilon^w$. Since ε^w is generally much less than 1, (5.45) represents that the player will a different strategy with a relatively smaller probability if its mood is 1 (i.e., $1 - \varepsilon^w \gg \frac{\varepsilon^w}{|\mathcal{S}_n|-1}$). The main motivation behind utilizing (5.44) and (5.45) to update $\mathbf{Q}_n(t)$ is that this rule guarantees that each individual player would more likely to choose the strategy making its mood be 1.

After that, player n will choose an action $\mathbf{s}_n(t)$ based on the probability distribution $\mathbf{Q}_n(t)$, calculate its utility $U_n(t)$ by measuring interference, and finally update mood $\alpha_n(t)$ with Algorithm 5.3.

Algorithm 5.3 Mood updating algorithm

1: **if** $\alpha_n (t - 1) = 1$ **then**
2:    **if** $(\mathbf{s}_n (t) = \mathbf{s}_n (t - 1))$ and $(U_n (t) = U_n (t - 1))$ **then**
3:       Set $\alpha_n (t)$ to 1
4:    **else**
5:       Go to 10.
6:    **end if**
7: **else**
8:    Go to 10.
9: **end if**
10: Set $\alpha_n (t)$ to 1 and 0 with the probability $\rho_1 = \varepsilon^{1 - U_n(t)}$ and $\rho_0 = 1 - \rho_1$, respectively.

Algorithm 5.4 UTC

1: Initialize iteration count $t = 0$, personality $\alpha_n (t) = 0$ and strategy counter $\mathbf{V}_n = \left(v_n^1, v_n^2, \cdots, v_n^{|\mathcal{S}_n|} \right) = (\mathbf{0})_{1 \times (|\mathcal{S}_n|)}$, $\forall n \in \mathcal{N}$. Each player n randomly chooses its initial strategy $\mathbf{s}_n(t)$ and then, measures its utility $U_n (t)$.
2: **repeat**
3:    Set $t = t + 1$
4:    **for** $n = 1$ to N users **do**
5:       Update state profile $\mathbb{L}_n (t)$:
6:       **if** $\alpha_n (t - 1) = 0$ **then**
7:          Calculate $\mathbf{Q}_n(t)$ with (5.44).
8:       **else**
9:          Calculate $\mathbf{Q}_n(t)$ with (5.45).
10:      **end if**
11:      Choose a strategy $\mathbf{s}_n (t)$, measure the utility $U_n (t)$, and update its mood $\alpha_n (t)$.
12:      Update strategies count $\mathbf{V}_n$:
13:      **if** $\alpha_n (t) = 1$ **then**
14:         Update $\mathbf{V}_n$ with (5.46).
15:      **end if**
16:    **end for**
17: **until** the stop criterion is satisfied.
18: Each player n decides its strategy $\mathbf{s}_n^D$ according to (5.47).

UTC Now, based on the above described state profile and learning model, UTC is developed and shown in Algorithm 5.4, where players can update their strategies in parallel. Similar to [35], the stop criterion of this algorithm can be one of the following: (1) the preset maximum iteration number T is reached or (2) for each player n, the variation of its utility during a period is trivial.

During the initialization of Algorithm 5.4, each player n will randomly choose its own strategy, set its moods to 0, and initialize the strategy counter $\mathbf{V}_n$, where

$(\mathbf{0})_{1\times(|\mathcal{S}_n|)}$ represents the $|\mathcal{S}_n|$-dimension null vector. We note that elements in vector $\mathbf{V}_n$ are used to count the times of $\alpha_n = 1$ when different strategies are adopted. For instance, v_n^i represents the times that the ith strategy makes the mood of player n be 1. When the initialization is completed, the algorithm goes into a loop, in which each individual player n will first update its state profile $\mathbb{L}_n(t) = (\mathbf{s}_n(t), U_n(t), \alpha_n(t))$ with the devised utility-based learning model at each iteration. We note that the SINR estimation can be done by sending a pilot or training sequence from the transmitter to receiver in practice [53]. Therefore, the utility can be measured by each autonomous user. Then, the strategy counter $\mathbf{V}_n = \left(v_n^1, v_n^2, \cdots, v_n^{|\mathcal{S}_n|}\right)$ is updated based on the current mood $\alpha_n(t)$. If $\alpha_n(t) = 1$,

$$v_n^{i(\mathbf{s}_n(t))} = v_n^{i(\mathbf{s}_n(t))} + 1, \forall \mathbf{s}_n(t) \in S_n, \tag{5.46}$$

where $v_n^{i(\mathbf{s}_n(t))}$ is the $i(\mathbf{s}_n(t))$th entry in vector $\mathbf{V}_n$. Intuitively, this updating rule implies that each player would like to record the strategy that makes its mood be 1. When the loop is exited, individual players will make their final decisions:

$$\mathbf{s}_n^D = \arg_{\mathbf{s}_n} \left(v_n^{i(\mathbf{s}_n)} = \max\left\{v_n^1, v_n^2, \cdots, v_n^{|\mathcal{S}_n|}\right\}\right), \forall n \in \mathcal{N}. \tag{5.47}$$

From (5.47), we note that the strategy recorded most frequently will be eventually adopted by users.

The reasons why we choose the above decision rule are two-fold. First, it only requires simple comparison operations when making final decision as shown in (5.47). Second, it can make the solution of problem $\mathbf{P}$ be asymptotically achieved under the given condition, which will be proved in the following section. We note that with the adopted learning model, system dynamics can be depicted as a perturbed Markov process and the parameter $\varepsilon > 0$ is the perturbation factor. Therefore, to show that the optimal strategy profile can be converged, it is essential to prove that the learning process of our algorithm will lead to a stochastically stable strategy profile that can maximize the overall utility. A similar idea has also been adopted by work [35] when designing the utility-based learning algorithm. However, authors in [35] adopted a quaternary variable instead of a binary variable to depict each user's mood, which introduces a much larger state space to capture system dynamics, and hence makes the convergence speed of their algorithm slower than that of ours. This will be illustrated through simulation results as shown in the following section.

Moreover, it is worth noting that Algorithm 5.4 is simple and completely distributed. In particular, when each player updates its own state profile, it does not require any prior information of other players, thereby avoiding a large communication overhead.

5.3.2.3 Performance Analysis of UTC

In this section, we first present the complexity analysis for the proposed algorithm UTC. Then, we will analyze its efficiency.

The main blocks of UTC are two parts. The first one is the loop from line 2 to line 17, which is independently executed by each player. The second one is the step in line 18, in which each player n needs to make its own final decision with (5.16). Note that the first main part (i.e., from line 2 to line 17) only involves basic arithmetic operations and random number generation, and hence has a computational complexity of $O(1)$ for each iteration. In addition, (5.16) requires the player n to compare the all $\mathcal{S}_n$ elements in the vector $\mathbf{V}_n$. Therefore, the complexity of this algorithm explicitly depends on both the stop criterion of the loop and the size of the player's strategy space. In particular, for the two different stop criterions described earlier, the complexities are $O(T+L)$ and $O(E+L)$, respectively, where T is the preset maximum iteration number, $L = \max\{|\mathcal{S}_1|, |\mathcal{S}_2|, \cdots, |\mathcal{S}_N|\}$, and E is the convergence speed of the algorithm. Moreover, it should be noted that the convergence speed E is related to the value of parameter ε, which will be further discussed at the end of this section.

Similar to the analysis of UDSA, we can show that UTC can also asymptotically converge to the solution maximizing the aggregate normalized utility under the given condition. Specifically, Let $\left(\mathbf{s}_n^O\right)_{n\in\mathcal{N}} \in (\mathcal{S}_n)_{n\in\mathcal{N}}$ denote the solution of problem $\mathbf{P}$; that is,

$$\left(\mathbf{s}_n^O\right)_{n\in\mathcal{N}} = \underset{(\mathbf{s}_n)_{n\in\mathcal{N}}}{\arg\ \max} \sum_{n\in\mathcal{N}} U_n\left((\mathbf{s}_n)_{n\in\mathcal{N}}\right). \tag{5.48}$$

When $\left(\mathbf{s}_n^O\right)_{n\in\mathcal{N}}$ is unique and ε is sufficiently small (i.e., $\varepsilon \to 0$), the solution of UTC asymptotically converges to $\left(\mathbf{s}_n^O\right)_{n\in\mathcal{N}}$; that is,

$$\Pr\left(\lim_{T\to\infty,\varepsilon\to 0}\left(\mathbf{s}_n^D\right)_{n\in\mathcal{N}} = \left(\mathbf{s}_n^O\right)_{n\in\mathcal{N}}\right) \doteq 1, \tag{5.49}$$

where T is the number of iterations.

We note that there is no requirement that the optimal solution $\left(\mathbf{s}_n^O\right)_{n\in\mathcal{N}}$ is a NE for the formulated game, and hence, this efficient point may be ignored by existing utility-based resource allocation algorithms that are designed to reach a NE [21, 34, 35]. In addition, it is worth noting that when the parameter ε is given, a much larger state space will make the convergence speed of the proposed algorithm much slower (i.e., there is a curse of dimensionality). This is mainly due to the fact that the considered resource allocation is essentially a combinatorial problem that is generally NP-hard. On the other hand, there is a trade-off between the efficiency and the convergence speed of our algorithm, which can be made by adjusting ε. Specifically, a smaller ε will lead to a slower convergence speed, but the algorithm is more likely to converge to the global optimal solution $\left(\mathbf{s}_n^O\right)_{n\in\mathcal{N}}$. For this reason, if ε is properly set then our algorithm still works in the scenario where the size of state space becomes large. In other words, a trade-off between the convergence speed and accuracy can be properly made to implement this algorithm in practice. This conclusion will be confirmed with simulation results in the following section.

Table 5.4: Simulation Parameters

Parameter	Value
Number of channels, K	5
Region radius, r	300 m
Maximum distance, D	20 m
Bandwidth per channel, B_0	0.1 MHz
AWGN power density, N_0	-174 dBm/Hz
Power level set, $\mathcal{P}_n$, $\forall n$	$\{-20, -10, 0\}$ dBW
Wall penetration loss, L_{wall}	5 dB
Steepness of the sigmoid function, β_n, $\forall n$	10
Path-loss exponent, α	3
Shadow fading standard deviation σ_ψ	4 dB

5.3.3 Results and Analysis

5.3.3.1 Simulation Scenario

To evaluate the performance of our proposed algorithm, we conduct simulations of a femtocell network consisting of N femtocells, which are randomly deployed in a circular region of radius r m. Meanwhile, the distance between each transmit-receive pair is a uniform random variable between 0 and D m. We assume that all the channels undergo identically and independently lognormal shadow fading as well as path loss and moreover, the path loss exponent α and the shadow fading standard deviation σ_ψ are set to 3 and 4 dB, respectively. We note that this channel model has been confirmed empirically to accurately model the variation in received power in some outdoor and indoor radio propagation environments (see e.g., [54] and references therein). In addition, the duration of a shadow fade lasts for multiple seconds or minutes, and hence changes at a much slower time scale [53].

We consider a three-level power set for each user; that is, low, medium, and high power levels, which are set to -20 dBW, -10 dBW, and 0 dBW, respectively. As well that, for each user n, the minimal rate requirement R_n^{min} and steepness of the sigmoid function ω_n are set to $\frac{1}{10}R_n^{max}$ and 10, respectively. In addition, each individual simulation result is obtained by averaging over 1,000 independent realizations of the users' locations and channel conditions. Unless specified otherwise, the simulation parameters are adopted as listed in Table 5.4.

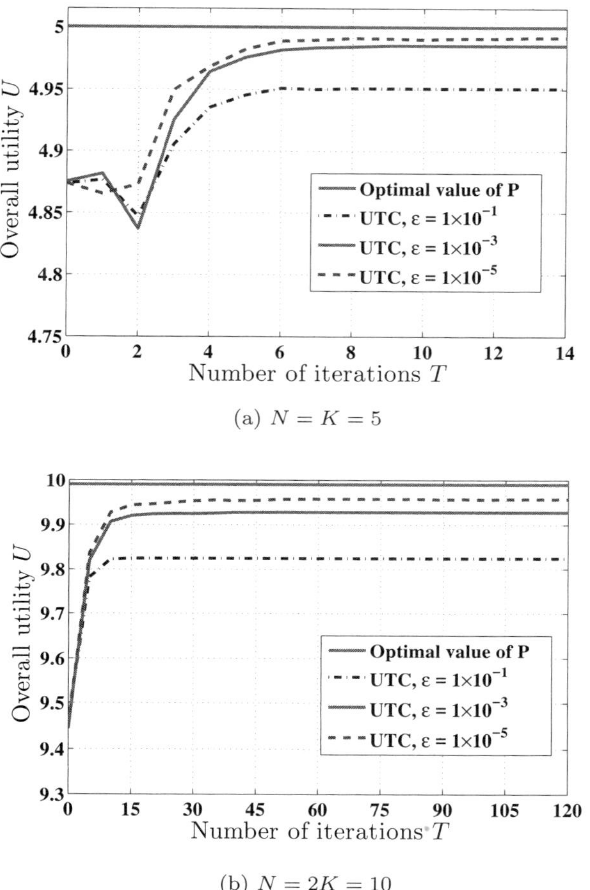

(a) $N = K = 5$

(b) $N = 2K = 10$

Figure 5.5: The convergence of UTC with respect to ε, where the number of channels is $K = 5$ and the overall utility is $U = \sum_{n \in N} U_n$.

5.3.3.2 Convergence of UTC

Before delving into the performance of the proposed distributed resource allocation algorithm UTC, we first investigate its convergence behavior and examine the impact of the algorithm parameter ε. We provide the maximum overall utility as shown in (5.37) as a benchmark result. In order to solve the problem $\mathbf{P}$ within an acceptable period of time, a simplified scenario is considered in this simulation. In particular, we focus on the case that there are $K = 5$ channels, and meanwhile, all the users transmit with the high power level (i.e., $\mathcal{P}_n = \{0 \text{ dBW}\}, \forall n \in \mathcal{N}$). When there are $N = K$ and $N = 2K$ users, the simulation results are illustrated in Figure 5.5(a) and 5.5(b), respectively.

It can be seen from Figure 5.5 that when ε becomes smaller, the convergence speed of UTC is slower but the achieved overall utility is higher. As well that, although there is a small gap between the performance of UTC and that of enumeration, our algorithm converges with much fewer iterations than that required by the latter (i.e., $\prod_{n \in \mathcal{N}} |\mathcal{S}_n| = \prod_{n \in \mathcal{N}} K |\mathcal{P}_n|$). Considering the scenario consisting of 10 users as an example, enumeration needs $5^{10} = 9765625$ iterations but our algorithm converges in about 40 and 100 iterations when ε is set to 10^{-3} and 10^{-5}, respectively. Moreover, if ε is set to 10^{-5}, the relative difference between the overall utility achieved by enumeration and that achieved with our algorithm is only around 0.4%. We note that this gap may stem from the fact that ε is not small enough and meanwhile there is no guarantee that the optimal solution of $\mathbf{P}$ is always unique in each round of simulation.

5.3.3.3   Performance Comparison

In this section we will evaluate the performance of our algorithm UTC with the following metrics:

- Overall utility U: The sum utility of the all players (i.e., $U = \sum_{n \in N} U_n$).

- Average transmission rate $\bar{R}$: The average transmission rate achieved by the users (i.e., $\bar{R} = \frac{\sum_{n \in \mathcal{N}} R_n}{N}$).

- User satisfaction ratio η_s: The ratio of users whose rate requirements are met over the total number of users N (i.e., $\eta_s = \frac{|\mathcal{N}_0|}{N}$, $\forall n \in \mathcal{N}_0, R_n \geq R_n^{min}$). Note that $|\cdot|$ denotes the cardinality of a set.

We compare our algorithm UTC with three distributed schemes that are presented as follows.

- Random: With this algorithm, each user n will randomly choose a strategy $\mathbf{s}_n$ from its strategy space $\mathcal{S}_n = \mathcal{K} \times \mathcal{P}_n$. Therefore, the performance of this method can be regarded as the baseline.

- Greedy transmission control (GTC): This greedy based algorithm is proposed in [55], with which each user needs to measure interference on all channels and then transmits on the channel having the minimum interference with the maximum transmit power. This process is repeated until the stop criterion is satisfied.

- TEL: This utility-based distributed learning algorithm is developed in [35]. Note that when implementing this algorithm, the corresponding mapping functions $G(x)$ and $F(x)$ are set the same as those adopted in previous Section 5.3.3.2.

When running UTC and TEL, we use the parameter setting suggested in [35], and hence set the perturbation factor ε to 10^{-2}. In addition, for fair comparison,

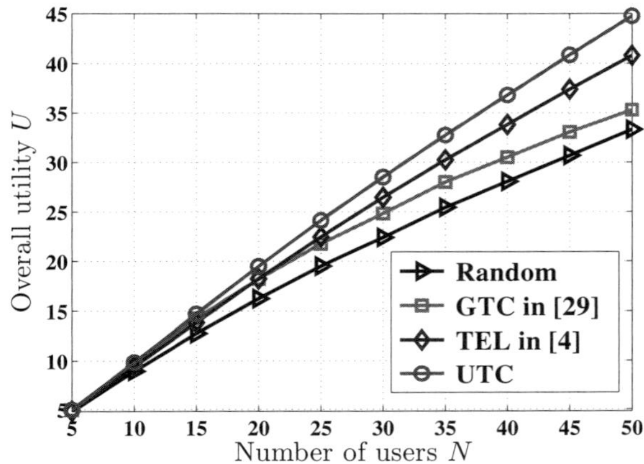

Figure 5.6: Overall utility U versus the number of users N.

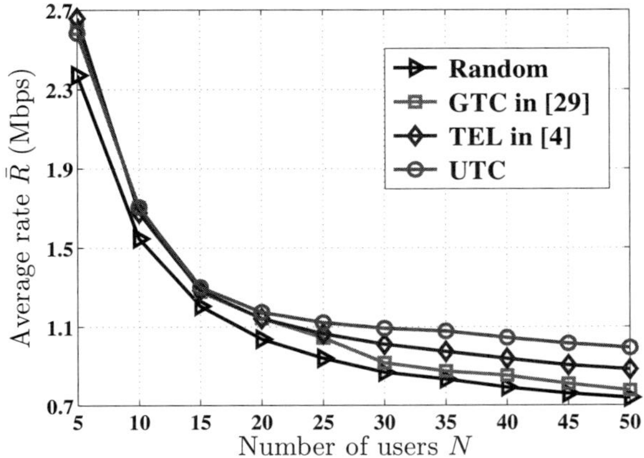

Figure 5.7: Average rate $\bar{R}$ versus the number of users N.

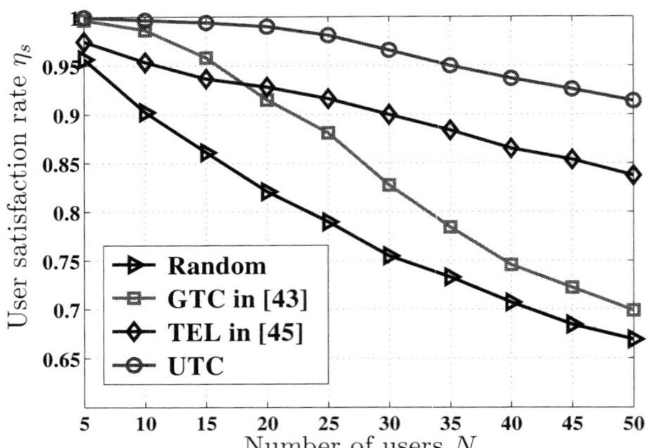

(a) User satisfaction rate η_s versus the number of users N.

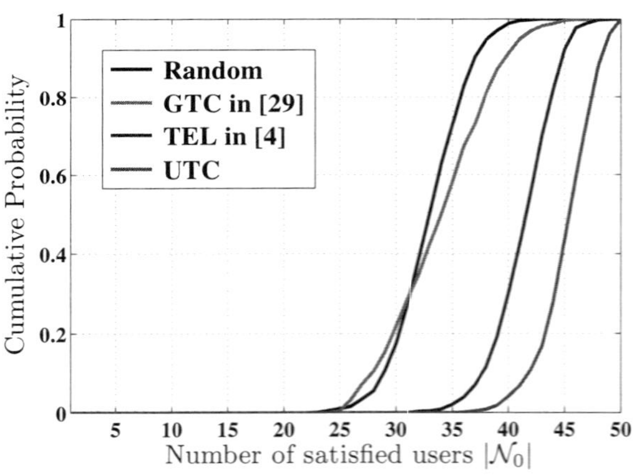

(b) The CDF of the number of satisfied users $|\mathcal{N}_0|$ for the algorithms, where there are $N = 50$ users.

Figure 5.8: Performance comparison in terms of satisfaction.

all the algorithms are executed in parallel and the maximum iteration number T is set to 10^4 [35].

Figures 5.6 and 5.7 illustrate the overall utility U and the corresponding average rate $\bar{R}$ versus the number of users, respectively. As can be observed from the simulation results, when there are more transmitting users, the improvement of the overall utility is decreased and meanwhile, the average rate becomes lower. This is because when the density of users increase there is more interference in this network, and in return, both the achieved utility and transmission rate of each user would decrease. In addition, we can see that when the number of communicating users in this network is small (for example $N \leq 15$), the performance of GTC is good. However, when there are more users its performance becomes worse. This is mainly caused by the greedy behavior of users when implementing this algorithm (i.e., they always transmit with the maximum power to improve their own utilities). As discussed in previous studies [20, 56], such a greedy-based method may cause severe interference in the system and finally may become an inefficient resource allocation strategy.

Additionally, we note that both our algorithm and TEL perform much better than the baseline algorithm (i.e., random). Meanwhile, compared with TEL, there is also an improvement in performance by implementing our algorithm. For instance, when there are $N = 50$ users, UTC has around 9.7% higher overall utility (i.e., from 40.08 to 44.77) and 12.4% higher average rate (i.e., from 0.884 Mbps to 0.994 Mbps) than TEL, respectively. Therefore, we can conclude that interference mitigation capability of UTC is the best among these four distributed algorithms. It should be noted that the reason for this improvement is similar to that stated in the previous section.

Next, we compare the performance of these four algorithms from the perspective of user satisfaction, which is demonstrated in Figure 5.8. In particular, Figure 5.8(a) illustrates the user satisfaction rate η_s versus the number of users N, and Figure 5.8(b) compares the cumulative distribution function (CDF) of the number of satisfied users (i.e., $|\mathcal{N}_0| = \eta_s \cdot N$) for the four algorithms when there are $N = 50$ users. Three observations can be made from Figure 5.8. First, not all the rate requirements of users can be met, especially for the case where the number of users is large. Second, η_s decreases with respect to N, which is due to the fact that more users will result in higher interference in this network. Third, compared with the other resource allocation schemes, more users can satisfy their rate requirements with our algorithm. Furthermore, combining the results shown in Figure 5.6, 5.7, and 5.8, we can see that our algorithm is able to achieve better performance on both the system and individual level. This is mainly because both the selfishness of users and welfare of the whole network are well considered when developing our algorithm.

5.4   Conclusions

In this chapter, we have addressed the issue of distributed subchannel allocation and power control for interference mitigation in OFDMA femtocells, and proposed a utility-based distributed resource allocation algorithm. The developed algorithm is appropriate for the networks that are organized in an ad hoc fashion, since there is no information interaction among the autonomous agents, and moreover, the Pareto-optimal resource allocation can be achieved under the given condition. Simulation results verify the validity of our analysis and demonstrate the effectiveness of our proposed scheme. Compared to the available strategies requiring information exchange, our approach achieves comparable or even better performance in different interference environments.

REFERENCES

[1] V. Chandrasekhar, J. Andrews, and A. Gatherer, "Femtocell networks: a survey," *IEEE Commun. Mag.*, vol. 46, no. 9, pp. 59–67, Sep. 2008.

[2] Y. Sun, R. Jover, and X. Wang, "Uplink interference mitigation for OFDMA femtocell networks," *IEEE Trans. Wireless Commun.*, vol. 11, no. 2, pp. 614–625, Feb. 2012.

[3] C.-X. Wang, F. Haider, X. Gao, X.-H. You, Y. Yang, D. Yuan, H. Aggoune, H. Haas, S. Fletcher, and E. Hepsaydir, "Cellular architecture and key technologies for 5G wireless communication networks," *IEEE Commun. Mag.*, vol. 52, no. 2, pp. 122–130, Feb. 2014.

[4] T. Zahir, K. Arshad, A. Nakata, and K. Moessner, "Interference management in femtocells," *IEEE Commun. Surveys Tuts.*, vol. 15, no. 1, pp. 293–311, Feb. 2013.

[5] N. Saquib, E. Hossain, L. B. Le, and D. I. Kim, "Interference management in OFDMA femtocell networks: issues and approaches," *IEEE Wireless Commun.*, vol. 19, no. 3, pp. 86–95, Jun. 2012.

[6] D. Lopez-Perez, A. Valcarce, G. de la Roche, and J. Zhang, "OFDMA femtocells: A roadmap on interference avoidance," *IEEE Commun. Mag.*, vol. 47, no. 9, pp. 41–48, Sep. 2009.

[7] D. Lopez-Perez, I. Guvenc, G. De la Roche, M. Kountouris, T. Quek, and J. Zhang, "Enhanced intercell interference coordination challenges in heterogeneous networks," *IEEE Wireless Commun.*, vol. 18, no. 3, pp. 22–30, Jun. 2011.

[8] L. Huang, G. Zhu, and X. Du, "Cognitive femtocell networks: an opportunistic spectrum access for future indoor wireless coverage," *IEEE Wireless Commun.*, vol. 20, no. 2, pp. 44–51, Apr. 2013.

[9] H.-C. Lee, D.-C. Oh, and Y.-H. Lee, "Mitigation of inter-femtocell interference with adaptive fractional frequency reuse," in *Proc. IEEE ICC'10*, Cape Town, South Africa, May 2010, pp. 1–5.

[10] Y.-S. Liang, W.-H. Chung, G.-K. Ni, I.-Y. Chen, H. Zhang, and S.-Y. Kuo, "Resource allocation with interference avoidance in OFDMA femtocell networks," *IEEE Trans. Veh. Technol.*, vol. 61, no. 5, pp. 2243–2255, Jun. 2012.

[11] R. Xie, F. Yu, H. Ji, and Y. Li, "Energy-efficient resource allocation for heterogeneous cognitive radio networks with femtocells," *IEEE Trans. Wireless Commun.*, vol. 11, no. 11, pp. 3910–3920, Nov. 2012.

[12] W. C. Cheung, T. Quek, and M. Kountouris, "Throughput optimization, spectrum allocation, and access control in two-tier femtocell networks," *IEEE J. Sel. Areas Commun.*, vol. 30, no. 3, pp. 561–574, Apr. 2012.

[13] D. T. Ngo, L. B. Le, T. Le-Ngoc, E. Hossain, and D. I. Kim, "Distributed interference management in two-tier CDMA femtocell networks," *IEEE Trans. Wireless Commun.*, vol. 11, no. 3, pp. 979–989, Mar. 2012.

[14] V. Chandrasekhar and J. Andrews, "Spectrum allocation in tiered cellular networks," *IEEE Trans. Wireless Commun.*, vol. 57, no. 10, pp. 3059–3068, Oct. 2009.

[15] P. Lee, T. Lee, J. Jeong, and J. Shin, "Interference management in LTE femtocell systems using fractional frequency reuse," in *Proc. International Conference on Advanced Communication Technology (ICACT)'10*, vol. 2, Gangwon-Do, Korea, Feb. 2010, pp. 1047–1051.

[16] D. Lopez-Perez, A. Ladanyi, A. Juttner, and J. Zhang, "OFDMA femtocells: A self-organizing approach for frequency assignment," in *Proc. IEEE PIMRC'09*, Tokyo, Japan, Sep. 2009, pp. 2202–2207.

[17] G. W. O. Costa, A. Cattoni, I. Kovacs, and P. Mogensen, "A fully distributed method for dynamic spectrum sharing in femtocells," in *Proc. IEEE Wireless Communications and Networking Conference Workshops (WCNCW)'12*, Paris, France, Apr. 2012, pp. 87–92.

[18] L. Garcia, K. Pedersen, and P. Mogensen, "Autonomous component carrier selection: interference management in local area environments for lte-advanced," *IEEE Commun. Mag.*, vol. 47, no. 9, pp. 110–116, 2009.

[19] W.-C. Hong and Z. Tsai, "Improving the autonomous component carrier selection for home enodebs in LTE-advanced," in *Proc. IEEE Consumer Communications and Networking Conference (CCNC)'11*, Las Vegas, NV, USA, Jan. 2011, pp. 627–631.

[20] A. MacKenzie and S. Wicker, "Game theory and the design of self-configuring, adaptive wireless networks," *IEEE Commun. Mag.*, vol. 39, no. 11, pp. 126–131, Nov. 2001.

[21] L. Rose, S. Lasaulce, S. Perlaza, and M. Debbah, "Learning equilibria with partial information in decentralized wireless networks," *IEEE Commun. Mag.*, vol. 49, no. 8, pp. 136–142, Aug. 2011.

[22] O. Aliu, A. Imran, M. Imran, and B. Evans, "A survey of self organisation in future cellular networks," *IEEE Commun. Surveys Tuts.*, vol. 15, no. 1, pp. 336–361, Feb. 2013.

[23] M. Peng, D. Liang, Y. Wei, J. Li, and H.-H. Chen, "Self-configuration and self-optimization in LTE-advanced heterogeneous networks," *IEEE Commun. Mag.*, vol. 51, no. 5, pp. 36–45, May 2013.

[24] A. Zappone, Z. Chong, E. Jorswieck, and S. Buzzi, "Energy-aware competitive power control in relay-assisted interference wireless networks," *IEEE Trans. Wireless Commun.*, vol. 12, no. 4, pp. 1860–1871, Apr. 2013.

[25] G. Bacci, E. V. Belmega, P. Mertikopoulos, and L. Sanguinetti, "Energy-aware competitive link adaptation in small-cell networks," in *Proc. International Workshop Resource Allocation in Wireless Networks'14*, Hammamet, Tunisia, May 2014.

[26] H. Kwon and B. G. Lee, "Distributed resource allocation through noncooperative game approach in multi-cell OFDMA systems," in *Proc. IEEE ICC'06*, vol. 9, Istanbul, Turkey, Jun. 2006, pp. 4345–4350.

[27] S. Buzzi, G. Colavolpe, D. Saturnino, and A. Zappone, "Potential games for energy-efficient power control and subcarrier allocation in uplink multicell OFDMA systems," *IEEE J. Sel. Topics Signal Process.*, vol. 6, no. 2, pp. 89–103, Apr. 2012.

[28] Q. D. La, Y. H. Chew, and B. H. Soong, "Performance analysis of downlink multi-cell OFDMA systems based on potential game," *IEEE Trans. Wireless Commun.*, vol. 11, no. 9, pp. 3358–3367, Sep. 2012.

[29] D. Monderer and L. Shapley, "Potential games," *Games and Economic Behavior*, vol. 14, no. 1, pp. 124–143, May 1996.

[30] R. Cominetti, E. Melo, and S. Sorin, "A payoff-based learning procedure and its application to traffic games," *Games and Economic Behavior*, vol. 70, no. 1, pp. 71–83, Sep. 2010.

[31] N. Li and J. Marden, "Designing games for distributed optimization," in *Proc. IEEE Conference on Decision and Control and European Control Conference (CDC-ECC)'11*, Orlando, Florida, USA, Dec. 2011, pp. 2434–2440.

[32] J. R. Marden, L. Y. Pao, and H. P. Young, "Achieving pareto optimality through distributed learning," Department of Economics, University of Oxford, Oxford, U.K., Tech. Rep., Jul. 2011.

[33] J. Marden, "State based potential games," *Automatica*, vol. 48, no. 12, pp. 3075 – 3088, Dec. 2012.

[34] Q. Wu, Y. Xu, J. Wang, L. Shen, J. Zheng, and A. Anpalagan, "Distributed channel selection in time-varying radio environment: Interference mitigation game with uncoupled stochastic learning," *IEEE Trans. Veh. Technol.*, vol. 62, no. 9, pp. 4524–4538, Nov. 2013.

[35] L. Rose, S. Perlaza, C. Le Martret, and M. Debbah, "Self-organization in decentralized networks: A trial and error learning approach," *IEEE Trans. Wireless Commun.*, vol. 13, no. 1, pp. 268–279, Jan. 2014.

[36] M. Bennis, S. Perlaza, P. Blasco, Z. Han, and H. Poor, "Self-organization in small cell networks: A reinforcement learning approach," *IEEE Trans. Wireless Commun.*, vol. 12, no. 7, pp. 3202–3212, Jul. 2013.

[37] P. S. Sastry, V. V. Phansalkar, and M. Thathachar, "Decentralized learning of Nash equilibria in multi-person stochastic games with incomplete information," *IEEE Trans. Syst., Man, Cybern.*, vol. 24, no. 5, pp. 769–777, May 1994.

[38] Z. Han, C. Pandana, and K. Liu, "Distributive opportunistic spectrum access for cognitive radio using correlated equilibrium and no-regret learning," in *Proc. IEEE WCNC'07*, Kowloon, 2007, pp. 11–15.

[39] V. Chandrasekhar, J. Andrews, T. Muharemovict, Z. Shen, and A. Gatherer, "Power control in two-tier femtocell networks," *IEEE Trans. Wireless Commun.*, vol. 8, no. 8, pp. 4316–4328, Aug. 2009.

[40] H. Claussen, "Performance of macro- and co-channel femtocells in a hierarchical cell structure," in *Proc. IEEE PIMRC'07*, Athens, Greece, Sep. 2007, pp. 1–5.

[41] O. Gharehshiran, A. Attar, and V. Krishnamurthy, "Collaborative subchannel allocation in cognitive LTE femto-cells: A cooperative game-theoretic approach," *IEEE Trans. Commun.*, vol. 61, no. 1, pp. 325–334, Jan. 2013.

[42] Q. D. La, Y. H. Chew, and B. H. Soong, "An interference-minimization potential game for OFDMA-based distributed spectrum sharing systems," *IEEE Trans. Veh. Technol.*, vol. 60, no. 7, pp. 3374–3385, Sep. 2011.

[43] T. Basar and G. J. Olsder, *The Theory of Learning in Games*. Cambridge, MA: MIT Press, 1999.

[44] Q. Wu, Y. Xu, J. Wang, L. Shen, J. Zheng, and A. Anpalagan, "Distributed channel selection in time-varying radio environment: Interference mitigation game with uncoupled stochastic learning," *IEEE Trans. Veh. Technol.*, vol. 62, no. 9, pp. 4524–4538, Nov. 2013.

[45] C. Xu, M. Sheng, X. Wang, C.-X. Wang, and J. Li, "Distributed subchannel allocation for interference mitigation in ofdma femtocells: A utility-based learning approach," *IEEE Trans. Veh. Technol.*, vol. 64, no. 6, pp. 2463–2475, Jun. 2015.

[46] I. Hwang, B. Song, and S. Soliman, "A holistic view on hyper-dense heterogeneous and small cell networks," *IEEE Commun. Mag.*, vol. 51, no. 6, pp. 20–27, Jun. 2013.

[47] M. Xiao, N. Shroff, and E. K. P. Chong, "A utility-based power-control scheme in wireless cellular systems," *IEEE/ACM Trans. Networking.*, vol. 11, no. 2, pp. 210–221, Apr. 2003.

[48] J. Zhang and Q. Zhang, "Stackelberg game for utility-based cooperative cognitiveradio networks," in *Proc. Proceedings of the Tenth ACM International Symposium on Mobile Ad Hoc Networking and Computing.* New York, NY, USA: ACM, 2009, pp. 23–32.

[49] H. Lin, M. Chatterjee, S. Das, and K. Basu, "ARC: an integrated admission and rate control framework for competitive wireless CDMA data networks using noncooperative games," *IEEE Trans. Mobile Computing*, vol. 4, no. 3, pp. 243–258, May 2005.

[50] D. T. Ngo, L. B. Le, T. Le-Ngoc, E. Hossain, and D. I. Kim, "Distributed interference management in two-tier CDMA femtocell networks," *IEEE Trans. Wireless Commun.*, vol. 11, no. 3, pp. 979–989, Mar. 2012.

[51] R. B. Myerson, *Game theory: analysis of conflict.* Cambridge, MA: Harvard university press, 2013.

[52] S. Buzzi, G. Colavolpe, D. Saturnino, and A. Zappone, "Potential games for energy-efficient power control and subcarrier allocation in uplink multicell OFDMA systems," *IEEE J. Sel. Topics Signal Process.*, vol. 6, no. 2, pp. 89–103, Apr. 2012.

[53] D. Tse and P. Viswanath, *Fundamentals of wireless communication.* Cambridge, UK: Cambridge university press, 2005.

[54] A. Goldsmith, *Wireless Communications.* Cambridge, UK: Cambridge University Press, 2004.

[55] B. Babadi and V. Tarokh, "Gadia: A greedy asynchronous distributed interference avoidance algorithm," *IEEE Trans. Inf. Theory*, vol. 56, no. 12, pp. 6228–6252, Dec. 2010.

[56] F. Wang, M. Krunz, and S. Cui, "Price-based spectrum management in cognitive radio networks," *IEEE J. Sel. Topics Signal Process.*, vol. 2, no. 1, pp. 74–87, Feb. 2008.

Chapter 6

Interference Alignment in MIMO Heterogeneous Networks

6.1 Introduction

Heterogeneous networks (HetNets), which consist of the conventional macrocell networks and a diverse set of overlaid small cells, are considered one of the most promising solutions to meet the explosively increasing traffic demand in wireless networks [1–7]. By deploying small cells, such as picocells and femtocells, in conventional cellular networks, the indoor and hotspots coverage of cellular networks is improved. Meanwhile, network capacity is significantly increased due to higher spatial spectrum reuse.

Although small cells can improve spectral efficiency, severe interference is also brought about by the deployment of small cells. Interference in HetNets can be categorized into two categories: cross-tier interference and cotier interference [2,3]. Cross-tier interference occurs among cells that belong to different tiers of HetNets, such as interference between macrocells and picocells. Cotier interference refers to interference among cells that belong to the same tier of HetNets, such as interference among picocells. These two types of interference significantly hinder the spatial reuse gain brought by small cells. Therefore, effective interference management is of vital importance to the success of HetNets.

In order to address interference problem in HetNets, lots of interference avoidance schemes, such as resource (frequency/time) partitioning approaches [4–7] are proposed. Due to the trend that BSs and user equipment will be equipped with multiple antennas, interference management by exploiting the multiple-input multiple-output (MIMO) technique has also been investigated [8–10]. Recently, an advanced beamforming technique, IA has been proposed to manage interference in wireless networks. The key idea of IA is to align interference from other transmitters in a reduced dimension subspace at each receiver [11]. Consequently, the remaining signal dimensions can be used for interference-free data transmission. It was shown that IA can achieve $K/2$ DoF in the K-user interference channel (i.e., the achievable DoF increases linearly with the number of users in the network). Inspired by the potential benefits of IA, many researchers have studied IA in cellular networks. As shown in [12–15], IA can effectively address interference in conventional macrocellular networks with MIMO antennas. However, the following characteristics of HetNets make it more challenging to apply IA in HetNets. The first is the heterogeneity of HetNets. The heterogeneous deployment of macro cells and small cells causes two types of interference (i.e., cross-tier interference and cotier interference), to the network. Due to the heterogeneity between the macro BSs and small cell BSs (i.e., the differences between the macro BSs and small cell BSs in terms of the transmission power and the number of equipped antennas), the cross-tier and cotier interference have different properties. As a result, designing IA schemes to address interference problem in HetNets is more challenging compared with conventional cellular networks. The second is the large number of small cells. Usually, there are a large number of small cells deployed in HetNets, while the numbers of antennas equipped by the BSs and user equipment are limited. Therefore, perfect IA is not feasible for the whole network [16–18].

There are several works that have applied IA to manage interference in Het-Nets [19–23]. In [19] and [20], IA is used to mitigate cross-tier interference. However, cotier interference among small cells is not effectively addressed. References [21, 22] focus on mitigating the cotier interference among small cells by dividing small cells into clusters that can achieve IA. Cotier interference in the same cluster is canceled by IA, but interference among clusters is not handled. Therefore, cotier interference among small cells is not fully addressed in these works. In [23], authors propose a hierarchical IA (HIA) scheme to address both the cross-tier and cotier interference in HetNets by exploiting the heterogeneity of HetNets. However, they only consider a scenario where IA is feasible. Due to the fact that BSs and user equipment are equipped with limited number of antennas, HIA is not achievable in HetNets when the number of small cells is large.

In above works, one of the most important characteristics of HetNets (i.e., partial connectivity[1] among small cells), is not exploited. As shown in [24, 25], partial connectivity can increase the DoF of interference channels. It is also shown in [26] that the performance of cellular networks can be improved by exploiting partial connectivity. However, due to the special characteristics of HetNets, the methods proposed in these works cannot be used in HetNets directly. This motivates us to design new schemes to exploit partial connectivity of HetNets.

---

[1]In a interference network, the signal strength between some receivers and interfering transmitters may be very weak due to large path loss, shadowing, and so forth. It is considered that there are no interference links between these transmitters and receivers because interference is negligible. This characteristic is called the partial connectivity of interference networks.

In this chapter, we focus on the downlink of a partially connected two-tier MIMO HetNet, which consists of one macrocell and several picocells. Specifically, we consider two partially connected HetNet scenarios. In the first scenario, we focus on partial connectivity among picocells, while in the second scenario partial connectivity between the macrocell and picocells is further considered. We first propose a two-stage IA scheme for the first scenario based on the discussion of IA feasibility conditions. By exploiting heterogeneity and partial connectivity structure of the considered HetNet, the proposed two-stage IA scheme increases the DoF of the HetNet. In HetNets, the macro BS is usually equipped with more antennas than pico BSs, and thus it can serve more users. However, due to cross-tier interference, the macro BS will cause interference to more signal dimensions of the picocell links when multiple macro users are served. Therefore, we further investigate how the number of served macro users affects the system DoF. Specifically, the condition under which serving one macro user achieves more DoF than serving multiple macro users are derived. Moreover, an algorithm is designed to determine the optimal number of served macro users so as to maximize the system DoF. Afterward, we study the second scenario, where some pico BSs are partially connected with macro users, and extend the two-stage IA scheme proposed above to this scenario.

IA can also be combined with other techniques. For example, interference alignment and cancellation (IAC) technique [27] is proposed by synthesizing IA and interference cancellation. In IAC, the receivers that have performed decoding operation will send the decoded data to the receivers that have not performed decoding operation for interference cancellation. As a result, interference caused by the decoded data can be canceled by reconstructing interference signals at the receivers that have not performed decoding operation. This significantly alleviates the requirement of IA in signal dimensions. Consequently, more DoF can be achieved by the IAC technique. However, the existing IAC schemes are all designed for homogeneous networks, such as wireless LANs [27] and conventional cellular networks [28, 29]. For HetNets, it is more challenging to design IAC schemes due to the randomly deployed small cells. For example, some small cells may suffer severe interference because a large number of small cells are deployed in their neighborhood, while some small cells suffer less interference due to the small cells being sparsely deployed. This characteristic brings several difficulties to the IAC scheme design for HetNets. First, it is hard to determine the optimal interference cancellation order so as to maximize the achievable system DoF. Second, for a given interference cancellation order, due to the limited antennas equipped by the BSs and user equipment, it necessary to determine how many data streams can be transmitted by each user and how to align them at each BS.

To address the above problems, in this chapter, we also propose an IAC scheme for the uplink of HetNets. First, in order to represent the complex interference relationship among cells, we model the transmission links in HetNets as a directed conflict graph. Then, based on the directed conflict graph we propose the IAC scheme, which comprises a set of constraints for users, BSs and interference cancellation, as well as the precoding and decoding design method for users and BSs. In the IAC scheme, we first develop the constraints for users and BSs so as to determine how IA should be performed for a given interference cancellation order. Then, the constraints on interference cancellation order are developed. Afterward, by giving the precoding and decoding design method, we prove that any IAC scheme that satisfies the developed constraints is feasible. In order to determine the optimal IAC scheme, we construct an optimization problem based on the developed constraints with the aim maximizing the number of data streams that can be transmitted in the network. By solving the optimization problem, we can determine how the optimal IAC scheme is performed in the HetNets. The simulation results show that the IAC scheme can significantly increase the number of data streams that can be transmitted in the network, and the achievable sum rate of the network is greatly increased.

6.2 Interference Alignment in Downlink Heterogeneous Networks

6.2.1 System Model

We consider the HetNet shown in Figure 6.1, which is composed of one macrocell and several picocells. The picocells are overlapped with the macrocell and all cells use the same frequency band. The macro BS is equipped with M_0 antennas and serves one or multiple users, while each pico BS has M antennas and serves one user. The users are equipped with N antennas. We assume that $M_0 \geq M \geq N$, because the macro BS usually has more antennas than pico BSs and BSs usually have more antennas than users. The picocells in Figure 6.1 can be categorized into two groups. The picocells in the first group are adjacent to each other and have strong mutual interference, such as the picocells represented with a solid line. The picocells in the second group (i.e., the picocells represented with a dasheded line), cause weak interference to each other and have weak mutual interference with the picocells in the first group. In practical HetNets, the picocells may not deployed as in Figure 6.1. Nevertheless, we can divide the picocells into several clusters through scheduling. Each cluster comprises the two picocell groups shown in Figure 6.1. Orthogonal resources are allocated to different clusters in order to avoid interference.

By neglecting the weak interference, the transmission links in the considered HetNet can be modeled as a partially connected interference network (PCIN), which is shown in Figure 6.2 (for simplicity, we only give the situation that one macro user is served). For convenience of representation, we divide the system into two subsystems. Subsystem 1 consists of the macrocell link and the picocell links in the first picocell group. Due to the high transmission power of the macro BS, we regard interference from the macro BS as strong interference at each picocell user. Meanwhile, with the aim of completely canceling the cross-tier interference, interference from pico BSs to the macro user is also treated as strong interference. As a result, the links in subsystem 1 form a fully connected interference network. The links in subsystem 2 correspond to the links in the

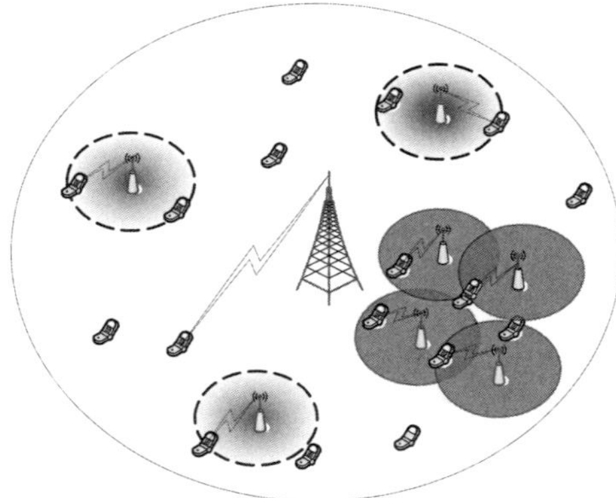

Figure 6.1: Illustration of the considered HetNet.

second pico cell group. Because the picocells in the second group cause weak interference to each other and have weak mutual interference with the picocells in the first group, each pico BS in subsystem 2 is only connected with its served user and the macro user. We suppose that the transmitters and receivers have perfect CSI.

We denote the BSs in subsystem 1 and subsystem 2 by $\mathcal{L} = \{0, 1, \ldots, L\}$ and $\mathcal{J} = \{L + 1, \ldots, L + J\}$, respectively. BS i transmits d_i data streams to its served user. For notational brevity, the user served by BS i is also denoted by i. Thus, the signals received at user i can be represented as

$$\boldsymbol{y}_i = \sqrt{g_{i,i}}\mathbf{H}_{i,i}\mathbf{V}_i\boldsymbol{x}_i + \sum_{j \in \mathcal{L} \cup \mathcal{J}} \sqrt{g_{i,j}}\mathbf{H}_{i,j}\mathbf{V}_j\boldsymbol{x}_j + \boldsymbol{n}_i, \forall i \in \mathcal{L} \cup \mathcal{J}, \qquad (6.1)$$

where $\boldsymbol{x}_i \in \mathbb{C}^{d_i \times 1}$ is the data vector transmitted by BS i, $\mathbf{V}_i \in \mathbb{C}^{M \times d_i}$ is the transmit precoding matrix of BS i, $\mathbf{H}_{i,j} \in \mathbb{C}^{N \times M}$ is the channel matrix from BS j to user i, $\boldsymbol{n}_i$ is the additive white Gaussian noise with zero mean and variance $\sigma^2 \mathbf{I}_N$, and $\sqrt{g_{i,j}}$ is the long-term path gain from BS j to user i. For the BSs and users that are disconnected with each other in Figure 6.2, the long-term path gains between them are very small. $\mathrm{E}\big[\|\boldsymbol{x}_i\|^2\big] = P_i$ is the transmission power of BS i. The first term and the second term of the right-hand side of (6.1) are the desired signal and interference signal received by user i, respectively. At the receiver side, user i decodes its desired signal through a decoding matrix $\mathbf{U}_i \in \mathbb{C}^{N \times d_i}$; that is,

$$\widetilde{\boldsymbol{y}}_i = \mathbf{U}_i^\dagger \boldsymbol{y}_i, \forall i \in \mathcal{L} \cup \mathcal{J}. \qquad (6.2)$$

The achievable rate of user i is given as follows

$$R_i = \log_2 \det \left(\mathbf{I}_{d_i} + \frac{g_{i,i}\mathbf{U}_i^\dagger\mathbf{H}_{i,i}\mathbf{V}_i\mathbf{P}_i\mathbf{V}_i^\dagger\mathbf{H}_{i,i}^\dagger\mathbf{U}_i}{A + B + \sigma^2\mathbf{I}_{d_i}} \right) \qquad (6.3)$$

where $A = \sum_{l \neq i, l \in \mathcal{L}} g_{i,l}\mathbf{U}_i^\dagger\mathbf{H}_{i,l}\mathbf{V}_l\mathbf{P}_l\mathbf{V}_l^\dagger\mathbf{H}_{i,l}^\dagger\mathbf{U}_i$ is interference caused by BSs in

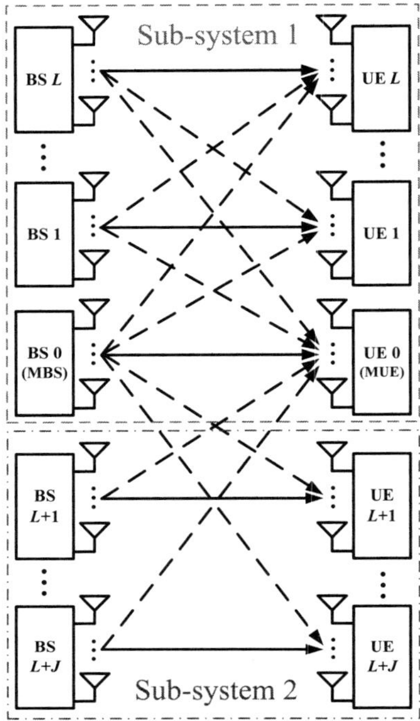

Figure 6.2: System model – PCIN (solid arrow: desired signal, dashed arrow: interference signal).

subsystem 1, $B = \sum_{j \neq i, j \in \mathcal{J}} g_{i,j} \times \mathbf{U}_i^\dagger \mathbf{H}_{i,j} \mathbf{V}_j \mathbf{P}_j \mathbf{V}_j^\dagger \mathbf{H}_{i,j}^\dagger \mathbf{U}_i$ is interference caused by BSs in subsystem 2, $\mathbf{P}_i = \mathrm{diag}\,(P_{i,1}, P_{i,2}, \ldots, P_{i,d_i})$ is the power allocation matrix of BS i, and $P_{i,j}$ is the power allocated to data stream j at BS i.

Notations: Bold uppercase letters denote matrices, such as $\mathbf{A}$, normal letters, such as a, denote scalars, and calligraphic letters, such as $\mathcal{A}$, denote sets. $\mathbf{I}_n$ represents the identity $n \times n$ matrix, $\mathbb{C}$ denotes complex field, $\mathbf{A}^\dagger$ is the conjugate transpose of $\mathbf{A}$. null($\mathbf{A}$) represents the nullspace of matrix $\mathbf{A}$, rank($\mathbf{A}$) denotes the rank of matrix $\mathbf{A}$. $(a)^+ = \max\{a, 0\}$. det($\mathbf{A}$) is the determinant of matrix $\mathbf{A}$. $\mathcal{A} \backslash \mathcal{B}$ represents the set of elements that are in $\mathcal{A}$ but not in $\mathcal{B}$. $\lfloor a \rfloor$ is the largest integer not greater than a, and $\lceil a \rceil$ denotes the smallest integer not less than a. $\mathrm{diag}(a_1, a_2, \ldots, a_n)$ represents a diagonal matrix with $a_1, a_2, \ldots, a_n$ as the diagonal elements.

6.2.2   Two-Stage IA

In this section, we first discuss about the IA feasibility conditions of the PCIN shown in Figure 6.2. Then, based on the discussion, a two-stage IA scheme is proposed. Afterward, we extend the two-stage IA to a scenario where multiple macro users are served. Moreover, the influence of the number of served macro users on the total system DoF is investigated.

6.2.2.1   IA Feasibility Conditions

In order to achieve IA, the following conditions must be satisfied by the PCIN;

$$\mathbf{U}_l^\dagger \mathbf{H}_{l,i} \mathbf{V}_i = \mathbf{0}, \forall i \neq l, i, l \in \mathcal{L}, \tag{6.4}$$

$$\mathbf{U}_0^\dagger \mathbf{H}_{0,j} \mathbf{V}_j = \mathbf{0}, \forall j \in \mathcal{J}, \tag{6.5}$$

$$\mathbf{U}_j^\dagger \mathbf{H}_{j,0} \mathbf{V}_0 = \mathbf{0}, \forall j \in \mathcal{J}, \tag{6.6}$$

$$\text{rank}\left(\mathbf{U}_i^\dagger \mathbf{H}_{i,i} \mathbf{V}_i\right) = d_i, \forall i \in \mathcal{L} \cup \mathcal{J}. \tag{6.7}$$

Condition (6.4) guarantees that at each user of subsystem 1 interference from other transmitters in subsystem 1 is aligned in a reduced subspace. Condition (6.5) assures that interference caused by transmitters in subsystem 2 will be aligned in interference subspace at the macro user, and condition (6.6) guarantees that the users of subsystem 2 are able to cancel interference from the macro BS. The last condition makes sure that the signal subspace $\mathbf{H}_{i,i} \mathbf{V}_i$ has dimension d_i and is linearly independent of interference subspace. We assume that all the elements of the channel matrices are randomly and independently generated from continuous distributions, thus condition (6.7) is automatically satisfied with probability 1 if we can find $\mathbf{V}_i$, $\mathbf{U}_i$ satisfying conditions (6.4)–(6.6) [30].

In the following, based on the above IA feasibility conditions, we investigate how many picocells can be accommodated into subsystem 1 when IA is feasible.

Lemma 1. Assume that all links of subsystem 1 convey the same number of data streams $d_i = d$, $i \in \mathcal{L}$, and M, N are divisible by d. If IA is feasible for the PCIN, the number of picocell links that can be accommodated into subsystem 1 is upper bounded by $(M + N)/d - 2$.

Proof. If IA is feasible for the PCIN, the links in subsystem 1 must be able to achieve IA (i.e., condition (6.4) must be satisfied). Note that the number of links in subsystem 1 is $L + 1$. Therefore, if $M_0 = M$, according to the IA feasibility conditions given by [17], IA is achievable for subsystem 1 if $M + N \geq (L + 2)d$ and M, N are divisible by d. Thus, the maximum number of picocell links that can be accommodated into subsystem 1 is $(M + N)/d - 2$.

When $M_0 > M$, IA is feasible for subsystem 1 if $L \leq (M + N)/d - 2$, because the macro BS only needs M antennas to achieve this upper bound. Now, we check whether IA is feasible when $L = (M + N)/d - 1$. According to the IA feasibility condition (6.4), the following equations must be satisfied if IA is feasible when $L = (M + N)/d - 1$

$$\mathbf{U}_l^\dagger \mathbf{H}_{l,i} \mathbf{V}_i = \mathbf{0}, \forall i \neq l, i, l \in \{1, 2, \ldots, L\}, \tag{6.8}$$

$$\mathbf{U}_0^\dagger \mathbf{H}_{0,i} \mathbf{V}_i = \mathbf{0}, \forall i \in \{1, 2, \ldots, L\}, \tag{6.9}$$

$$\mathbf{U}_i^\dagger \mathbf{H}_{i,0} \mathbf{V}_0 = \mathbf{0}, \forall i \in \{1, 2, \ldots, L\}. \tag{6.10}$$

According to [16, 17], if IA is feasible, for all subsets of equations, the number of variables involved cannot be less than the number of equations in that subset. We consider an equation subset contains (6.8) and (6.9). By eliminating superfluous variables, the total number of variables contained by (6.8) and (6.9) is $L(M + N - 2d)d + (N - d)d$, while the total number of equations contained by (6.8) and (6.9) is $L^2 d^2$. Thus, the following equation must hold if IA is feasible

$$L(M + N - 2d)d + (N - d)d \geq L^2 d^2. \tag{6.11}$$

When $L = (M + N)/d - 1$,

$$L(M + N - 2d)d + (N - d)d - L^2 d^2 = -Md < 0. \tag{6.12}$$

Therefore, the condition in (6.11) cannot hold. This indicates that IA is not feasible if $L = (M + N)/d - 1$. Therefore, the maximum number of picocell links that can be accommodated into subsystem 1 is $(M + N)/d - 2$. $\quad\square$

Lemma 1 tells us that even if the macro BS has more antennas, the number of picocell links that can be accommodated into subsystem 1 is still upper bounded by $(M + N)/d - 2$, and the macro BS only needs M antennas to achieve this upper bound.

Now, we study subsystem 2. From Figure 6.2 we can see that links in subsystem 2 have mutual interference only with the macrocell link. Thus, the macrocell link has a significant effect on the achievable DoF of subsystem 2. In the following, we investigate the DoF that can be achieved by each link in subsystem 2 for the given $\mathbf{U}_0$ and $\mathbf{V}_0$.

Lemma 2. For the given $\mathbf{U}_0$ and $\mathbf{V}_0$, the number of data streams d' that can be conveyed by each link in subsystem 2 is upper bounded by

$$d' \leq \min\{M - d, N - d\}. \tag{6.13}$$

where d is the number of data streams transmitted by the macro BS.

Proof. In order to achieve IA, conditions (6.5) and (6.6) must be satisfied simultaneously for each link in subsystem 2. When $\mathbf{U}_0$ and $\mathbf{V}_0$ are given, for each link in subsystem 1, the numbers of equations contained by both (6.5) and (6.6) are dd', and the numbers of variables involved in (6.5) and (6.6) are $(M - d')d'$ and $(N - d')d'$, respectively. Therefore, d' must satisfy the following conditions;

$$\begin{cases} (M - d')d' \geq dd' \\ (N - d')d' \geq dd' \end{cases}$$

$$\Rightarrow \begin{cases} d' \leq M - d \\ d' \leq N - d \end{cases}$$

$$\Rightarrow d' \leq \min\{M - d, N - d\}.$$

$\quad\square$

Lemma 2 shows that the number of data streams that can be conveyed by each link in subsystem 2 is determined by the number of data streams transmitted by the macro BS. Note that in order to achieve IA, the macrocell link must sacrifice some signal dimensions to accommodate interference signals, thus it only transmits data on parts of signal dimensions. This enables the links in subsystem 2 to obtain transmission opportunity.

6.2.2.2 Design of Two-Stage IA

Based on the above discussion, we propose a two-stage IA scheme that exploits the heterogeneity and partial connectivity of HetNets. In the following, we describe the main idea of two-stage IA briefly, and then the design details are elaborated. In the first stage, we first make use of the redundant antennas of the macro BS to cancel interference to parts or all of users in subsystem 2. This goal is fulfilled by enabling the macro BS to transmit on the nullspace of the channel matrices from the macro BS to some users in subsystem 2. Then, a conventional IA scheme is carried out to calculate the precoding and decoding matrices for links in subsystem 1. In the second stage, we design the precoding and decoding matrices for links in subsystem 2 by exploiting the partial connectivity of picocells. Due to partial connectivity, the picocell links in subsystem 2 do not need to address the cotier interference. Thus, only the cross-tier interference needs to be solved. For the links with the interference from the macro BS being canceled, the precoding and decoding matrices are designed to cancel interference to the macro user and maximize the achievable rates. For the links that are interfered by the macro BS, we design the precoding and decoding matrices with the goal of completely canceling the cross-tier interference and maximize their achievable rates by exploiting the signal dimensions that are not interfered by the macro BS. In the following, the details of the two-stage IA scheme are presented.

First stage IA When the macro BS has redundant antennas, we can enable the macro BS to transmit on the nullspace of the channel matrices from the macro BS to some users in subsystem 2. As a result, interference from the macro BS to these users is canceled. To this end, the macro BS uses a concatenated precoding scheme (i.e., $\mathbf{V}_0 = \mathbf{W}_0 \widetilde{\mathbf{V}}_0$). In particular, $\mathbf{W}_0$ cancels interference from the macro BS to some users in subsystem 2, while $\widetilde{\mathbf{V}}_0$ is designed to achieve IA for subsystem 1. $\mathbf{W}_0$ is given as follows

$$\mathbf{W}_0 = \mathrm{null}\left(\left[\mathbf{H}_{L+1,0}^\dagger, \ldots, \mathbf{H}_{L+q,0}^\dagger\right]^\dagger\right), \tag{6.14}$$

where $q = \min\left\{\left\lfloor \frac{M_0 - M}{N} \right\rfloor, J\right\}$ is the maximum number of links in subsystem 2 with interference from the macro BS being canceled. By using the precoding matrix $\mathbf{W}_0$, the macro BS will not cause interference to links in $\hat{\mathcal{J}} = \{L+1, L+2, \ldots, L+q\} \subseteq \mathcal{J}$. Afterward, the precoding and decoding matrices of links in subsystem 1 are obtained by carrying out a conventional IA scheme. In this process, equivalent channel matrices $\widetilde{\mathbf{H}}_{l,0} = \mathbf{H}_{l,0}\mathbf{W}_0, l \in \mathcal{L}$ are used. Note that closed-form IA solutions for fully connected interference networks are known only for certain cases [11, 31], iterative algorithms [30, 32] may be needed when subsystem 1 performs the conventional IA.

Second stage IA　In the following, the design of precoding and decoding matrices for the picocell links in subsystem 2 is given. In order to avoid interfering the macro user, the pico BSs in subsystem 2 should align their transmitted signals in interference subspace of the macro user (i.e., the subspace spanned by the columns of $\mathbf{G}_j \in \mathbb{C}^{M \times (M-d)}$). Specifically,

$$\mathbf{G}_j = \text{null}\left(\mathbf{U}_0^\dagger \mathbf{H}_{0,j}\right), j \in \mathcal{J}. \tag{6.15}$$

Therefore, the transmit precoding matrices of links in $\mathcal{J}$ can be represented as

$$\mathbf{V}_j = \mathbf{G}_j \widetilde{\mathbf{V}}_j, j \in \mathcal{J}, \tag{6.16}$$

where $\widetilde{\mathbf{V}}_j$ is a matrix that combines the column space of $\mathbf{G}_j$.

For links in $\hat{\mathcal{J}}$, they are not interfered by the macro BS. Thus, each link in $\hat{\mathcal{J}}$ can be regarded as a MIMO point-to-point channel with the following equivalent channel

$$\widetilde{\mathbf{H}}_{j,j} = \mathbf{H}_{j,j}\mathbf{G}_j, j \in \hat{\mathcal{J}}. \tag{6.17}$$

Then, the precoding and decoding matrices can be obtained by performing singular value decomposition (SVD) on channel matrix $\widetilde{\mathbf{H}}_{j,j}$ [33]. The SVD of $\widetilde{\mathbf{H}}_{j,j}$ is

$$\widetilde{\mathbf{H}}_{j,j} = \widetilde{\mathbf{U}}_j \mathbf{\Lambda}_j \widetilde{\mathbf{V}}_j^\dagger, j \in \hat{\mathcal{J}}. \tag{6.18}$$

Thus, $\mathbf{U}_j = \widetilde{\mathbf{U}}_j$ and $\mathbf{V}_j = \mathbf{G}_j \widetilde{\mathbf{V}}_j$ are the desired receive and transmit precoding matrices of link j, respectively. In order to improve the achievable rate, the water-filling power allocation over the singular values of $\widetilde{\mathbf{H}}_{j,j}$ is used; that is,

$$P_{j,n} = \left(\mu - \frac{\sigma^2}{\lambda_n^2}\right)^+, \tag{6.19}$$

where $n = 1, \ldots, \min\{M - d, N\}$, and λ_n are the diagonal elements of $\mathbf{\Lambda}_j$. μ is chosen to satisfy the total transmission power constraint $\sum_n P_{j,n} = P_j$.

For links in $\mathcal{J}\backslash\hat{\mathcal{J}}$, interference from the macro BS has to be canceled by the users. Therefore, the users should receive the desired signal in the subspace spanned by columns of

$$\mathbf{F}_j = \text{null}\left(\left(\mathbf{H}_{j,0}\mathbf{V}_0\right)^\dagger\right), j \in \mathcal{J}\backslash\hat{\mathcal{J}}, \tag{6.20}$$

where $\mathbf{F}_j \in \mathbb{C}^{N \times (N-d)}$ is the orthogonal complement of the received signal from the macro BS. Therefore, the actual decoding matrix of user $j \in \mathcal{J}\backslash\hat{\mathcal{J}}$ can be represented as

$$\mathbf{U}_j = \mathbf{F}_j \widetilde{\mathbf{U}}_j, j \in \mathcal{J}\backslash\hat{\mathcal{J}}. \tag{6.21}$$

The precoding and decoding matrices of link $j \in \mathcal{J} \backslash \hat{\mathcal{J}}$ can be obtained by performing SVD on the equivalent direct channel matrix

$$\widetilde{\mathbf{H}}_{j,j} = \mathbf{F}_j^\dagger \mathbf{H}_{j,j} \mathbf{G}_j, j \in \mathcal{J} \backslash \hat{\mathcal{J}}. \tag{6.22}$$

Thus, $\mathbf{V}_j = \mathbf{G}_j \widetilde{\mathbf{V}}_j$ and $\mathbf{U}_j = \mathbf{F}_j \widetilde{\mathbf{U}}_j$ are the desired precoding and decoding matrices, respectively. Also, the water-filling power allocation over the singular values is used to improve the achievable rate.

The links in $\hat{\mathcal{J}}$ can convey $\min\{M - d, N\}$ data streams, because only condition (6.5) needs to be satisfied by these links. For links in $\mathcal{J} \backslash \hat{\mathcal{J}}$, both conditions (6.5) and (6.6) should be satisfied. Thus, they can only convey $\min\{M - d, N - d\}$ data streams. Note that the DoF achieved by subsystem 1 is $M + N - d$. Thus, the total DoF can be achieved by the two-stage IA is

$$D = M + N - d + q \min\{M - d, N\} + (J - q)(N - d). \tag{6.23}$$

6.2.2.3 One Macro User or Multiple Macro Users?

In the above work, we investigate the scenario where only one user is served by the macro BS. As the macro BS usually has more antennas than pico BSs, it can serve more users and accordingly increase the DoF of subsystem 1. However, the macro BS will cause interference to more signal dimensions of picocell links in subsystem 2 when multiple macro users are served. Therefore, the number of served macro users has a significant impact on the system DoF.

In this subsection, the influence of the number of served macro users on the system DoF is investigated. To this end, we consider a scenario where the macro BS serves multiple users and each pico BS serves one user. Then, we extend the two-stage IA scheme proposed in the above section to this scenario. Based on the proposed scheme, we derive the condition under which the system can achieve more DoF by serving one macro user than by serving multiple macro users. Furthermore, an algorithm is designed to determine the optimal number of served macro users such that maximum system DoF can be achieved.

We assume that S users are served by the macro BS. In this scenario, we employ the generalized HIA (refers to GHIA) scheme proposed in [23] to calculate the precoding and decoding matrices of links in subsystem 1. In the following, we describe the main idea of GHIA briefly. The precoding and decoding matrices design of GHIA is divided into two steps. In the first step, the precoding and decoding matrices of picocell links and the decoding matrices of the macro users are designed with the aim to align interference from the pico BSs at each macro user and each picocell user. By denoting the S users served by the macro BS as $\mathcal{S} = \{k_1, \ldots, k_S\}$, the following conditions must be satisfied by the first step of GHIA [23];

$$\mathbf{U}_i^\dagger \mathbf{H}_{i,j} \mathbf{V}_j = 0, \forall i \neq j, i, j \in \{1, \ldots, L\}, \tag{6.24}$$

$$\mathbf{U}_{k_s}^\dagger \mathbf{H}_{k_s,j} \mathbf{V}_j = 0, \forall j \in \{1, \ldots, L\} \text{ and}$$

$$\forall s = 1, 2, \ldots, S, \tag{6.25}$$

$$\mathrm{rank}\left(\mathbf{U}_i^\dagger \mathbf{H}_{i,i} \mathbf{V}_i\right) = d, \forall i \in \{1, \ldots, L\}. \tag{6.26}$$

Condition (6.24) guarantees that IA is achieved among picocell links, condition (6.25) makes sure that interference from pico BSs is aligned at all macro users, and condition (6.26) assures that each user can obtain d DoF.

In the second step, the precoding matrix of the macro BS is designed based on the idea of zero-forcing precoding. The precoding matrix makes sure that interference caused by the macro BS will lie in interference subspace of each user. The feasibility conditions corresponding to the first and second steps are given by (6.27) and (6.28) [23], respectively.

$$M + \kappa N \geq \kappa \left(L + 1\right) d, \tag{6.27}$$

$$M_0 \geq \kappa L d, \tag{6.28}$$

where $\kappa = \frac{S}{L} + 1$.

When the macro BS serves multiple users, we modify the two-stage IA scheme proposed in Section 6.2.2.2 to this scenario. In particular, in the first stage, GHIA is used to calculate the precoding and decoding matrices for links in subsystem 1 instead of the conventional IA. Denote the maximum number of picocell links that can be accommodated into subsystem 1 by L_S when S macro users are served. Because at least $(S + L_S) d$ signal dimensions are needed by GHIA as shown in (6.28), the macro BS can cancel interference to $q' = \min\left\{\left\lfloor \frac{M_0 - (S + L_S)d}{N} \right\rfloor, J\right\}$ links in subsystem 2 by employing $\overline{\mathbf{W}}_0$, which is given by

$$\overline{\mathbf{W}}_0 = \mathrm{null}\left(\left[\mathbf{H}_{L+1,0}^\dagger, \ldots, \mathbf{H}_{L+q',0}^\dagger\right]^\dagger\right). \tag{6.29}$$

Then, links in subsystem 1 perform GHIA to obtain the precoding and decoding matrices. In this process, the equivalent channel matrices $\widetilde{\mathbf{H}}_{l,0} = \mathbf{H}_{l,0}\overline{\mathbf{W}}_0, l \in \mathcal{L} \backslash \{0\}$ and $\widetilde{\mathbf{H}}_{k_s,0} = \mathbf{H}_{k_s,0}\overline{\mathbf{W}}_0, s \in \mathcal{S}$ are used. We denote by $\widehat{\mathcal{J}}_S = \{L + 1, \ldots, L + q'\}$ the q' links in subsystem 2 that are not interfered by the macro BS.

In the second stage, the pico BSs in subsystem 2 should align the transmitted signals in interference subspaces of all macro users. In other words, the signal transmitted by BS $j \in \mathcal{J}$ must lie in the space spanned by the columns of $\overline{\mathbf{G}}_j$, which is given by

$$\overline{\mathbf{G}}_j = \mathrm{null}\left(\left[\left(\mathbf{U}_{k_1}^\dagger \mathbf{H}_{k_1,j}\right)^\dagger, \ldots, \left(\mathbf{U}_{k_S}^\dagger \mathbf{H}_{k_S,j}\right)^\dagger\right]^\dagger\right), j \in \mathcal{J}. \tag{6.30}$$

The size of matrix $[(\mathbf{U}_{k_1}^\dagger \mathbf{H}_{k_1,j})^\dagger,\ldots,(\mathbf{U}_{k_S}^\dagger \mathbf{H}_{k_S,j})^\dagger]^\dagger$ is $Sd \times M$. The condition that this matrix has a nullspace is $M > Sd$. If $M \leq Sd$, the links in subsystem 2 cannot transmit because they cannot cancel their interference to macro users.

For the links in $\widehat{\mathcal{J}}_S$, the precoding and decoding matrices are obtained by performing SVD on the equivalent direct channel matrix

$$\widetilde{\mathbf{H}}_{j,j} = \mathbf{H}_{j,j}\overline{\mathbf{G}}_j, j \in \widehat{\mathcal{J}}_S. \tag{6.31}$$

Thus, the number of data streams that can be conveyed by each link in $\widehat{\mathcal{J}}_S$ is $(\min\{M - Sd, N\})^+$.

For the links in $\mathcal{J}\backslash\widehat{\mathcal{J}}_S$, users need to cancel interference from the macro BS. Therefore, the subspace that is not interfered by the macro BS is

$$\overline{\mathbf{F}}_j = \mathrm{null}\left([\mathbf{H}_{j,k_1}\mathbf{V}_{k_1},\ldots,\mathbf{H}_{j,k_S}\mathbf{V}_{k_S}]^\dagger\right), j \in \mathcal{J}\backslash\widehat{\mathcal{J}}_S, \tag{6.32}$$

where matrix $[\mathbf{H}_{j,k_1}\mathbf{V}_{k_1},\ldots,\mathbf{H}_{j,k_S}\mathbf{V}_{k_S}]^\dagger$ is of size $Sd \times N$. If $N \leq Sd$, users in $\mathcal{J}\backslash\widehat{\mathcal{J}}_S$ cannot cancel interference from the macro BS. As a result, these links cannot transmit. Therefore, the condition that the links in $\mathcal{J}\backslash\widehat{\mathcal{J}}_S$ can transmit is $\min\{M, N\} > Sd$, and the maximum number of data streams that can be conveyed by each link is $(\min\{M - Sd, N - Sd\})^+$. The precoding and decoding matrices of these links are obtained by performing SVD on the equivalent direct channel matrix

$$\widetilde{\mathbf{H}}_{j,j} = \overline{\mathbf{F}}_j^\dagger\mathbf{H}_{j,j}\overline{\mathbf{G}}_j, j \in \mathcal{J}\backslash\widehat{\mathcal{J}}_S. \tag{6.33}$$

Similarly to Section 6.2.2.2, the water-filling power allocation is used to improve the achievable rate.

The achievable DoF of the system when S macro users are served is

$$D' = (S + L_S)\,d + q'\,(\min\{M - Sd, N\})^+$$
$$+ (J - q')\,(N - Sd)^+, \tag{6.34}$$

where the first term is the DoF achieved by subsystem 1, while the second and the third terms represent the DoF achieved by links in $\widehat{\mathcal{J}}_S$ and $\mathcal{J}\backslash\widehat{\mathcal{J}}_S$, respectively.

In the following, we give the condition under which the system achieves more DoF when one macro user is served in Theorem 1. Before presenting Theorem 1, we first give the following lemma.

Lemma 3. Assume that $M_0 \geq M \geq N$ and M, N are divisible by d. If S users are served by the macro BS, the following conclusions hold:

(1) $L_S \leq (M + N)/d - 2$;
(2) $(S + L_S)\,d \geq M + N - d$ if $M_0 \geq M + N - d$;
(3) $q \geq q'$.

Proof. (1) Because the signals from pico BSs must be aligned at each macro user, the more users are served by the macro BS, the more signal dimensions must be sacrificed by the picocell links so that the transmitted signals can be aligned at the macro users. As a result, less signal dimensions are left for IA construction among picocell links. Therefore, when $S = 1$ the maximum number of picocell links in subsystem 1 can be achieved. From Lemma 1, we know that the maximum number of pico cell links can be accommodated into subsystem 1 is $(M + N)/d - 2$. Therefore, L_S cannot exceed $(M + N)/d - 2$.

(2) In the following, we show that if $M_0 \geq M + N - d$, subsystem 1 can achieve more DoF by serving multiple macro users than by serving one macro user.

We rewrite (6.27) and (6.28) as

$$S(Ld + d - N) \leq L(M + N - Ld - d), \tag{6.35}$$

$$(L + S)d \leq M_0. \tag{6.36}$$

From (6.35), we have

$$L^2 d - L(M + N - Sd - d) - S(N - d) \leq 0. \tag{6.37}$$

Therefore, we can obtain that

$$L \leq \left\lfloor \frac{\Phi + \sqrt{\Phi^2 + 4dS(N - d)}}{2d} \right\rfloor, \tag{6.38}$$

where $\Phi = M + N - (S + 1)d$. Note that L is also constrained by (6.36), thus L_S can be expressed as

$$L_S = \min \left\{ \left\lfloor \frac{\Phi + \sqrt{\Phi^2 + 4Sd(N - d)}}{2d} \right\rfloor, \left\lfloor \frac{M_0}{d} \right\rfloor - S \right\}. \tag{6.39}$$

If $L_S = \left\lfloor \frac{\Phi + \sqrt{\Phi^2 + 4Sd(N-d)}}{2d} \right\rfloor$, due to the fact that $4Sd(N - d) \geq 0$ and M, N are divisible by d, then we have

$$L_S \geq \left\lfloor \frac{\Phi}{d} \right\rfloor = \frac{M + N}{d} - (S + 1). \tag{6.40}$$

Therefore, $(S + L_S)d \geq M + N - d$.

If $L_S = \left\lfloor \frac{M_0}{d} \right\rfloor - S$, then we can obtain that

$$(S + L_S)d = \left\lfloor \frac{M_0}{d} \right\rfloor d. \tag{6.41}$$

From (6.40) and (6.41), we can see that if $M_0 \geq M + N - d$, then $(S + L_S)d \geq M + N - d$. That is, more DoF can be achieved by subsystem 1 if the macro BS serves multiple macro users.

(3) From (6.40) and (6.41), we have

$$(S + L_S)d \geq \min \left\{ M + N - d, \left\lfloor \frac{M_0}{d} \right\rfloor d \right\}. \tag{6.42}$$

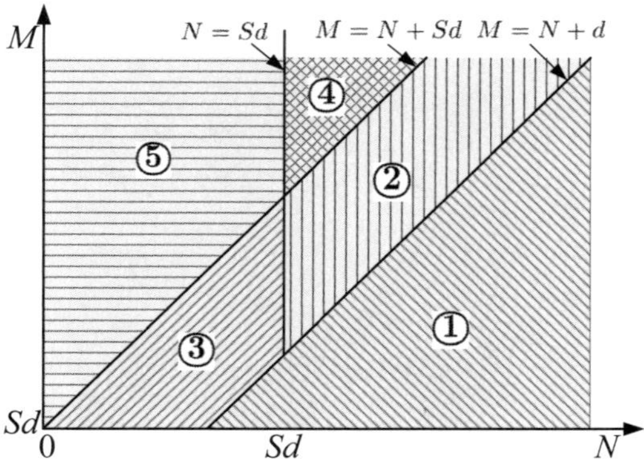

Figure 6.3: Regions for different cases.

Because we assume that $M_0 \geq M$ and M is divisible by d, we can obtain that $(S + L_S)\,d \geq M$. Thus, according to the definitions of q and q', we can obtain that $q \geq q'$. $\qquad\square$

Based on Lemma 3, we have the following theorem.

Theorem 2. Under the condition that $q < J$, the system can achieve more DoF when one macro user is served than when $S\,(2 \leq S < M/d)$ macro users are served.

Proof. In the following, we compare the DoF achieved by the system when one macro user and $S(2 \leq S < M/d)$ macro users are served. From (6.23) we can see that the value of D is closely related to the relationship between $M - d$ and N. Similarly, the value of D' in (6.34) is determined by the relationship among M, N, and Sd. Thus, we compare the values of D and D' under all cases. When $S < M/d$, we should compare the values of D and D' in five cases, which are shown in Figure 6.3.

- **Case 1:** $0 < M - Sd < N,\ M - d \leq N$

In this case, D and D' can be expressed as

$$D = M + N - d + q\,(M - d) + (J - q)\,(N - d), \tag{6.43}$$

$$D' = (S + L_S)\,d + q'\,(M - Sd) + (J - q')\,(N - Sd)^{+}. \tag{6.44}$$

Due to $q \geq q'$ and $M \geq N$, we have

$$D' \leq (S + L_S)\,d + q\,(M - Sd) + (J - q)\,(N - Sd)^{+}. \tag{6.45}$$

Thus, we can obtain that

$$
\begin{aligned}
D - D' \\
\geq M + N - d + q\left(M - d\right) - \left(S + L_S\right)d - q\left(M - Sd\right) \\
= \left(M + N - d - L_S d\right) + q\left(S - 1\right)d - Sd \\
\overset{(a)}{\geq} d + q\left(S - 1\right)d - Sd \\
= \left(q - 1\right)\left(S - 1\right)d,
\end{aligned}
\tag{6.46}
$$

where inequality (a) follows from conclusion (1) of Lemma 3. Therefore, if $q \geq 1$, then $D - D' \geq 0$.

If $q = 0$, then it follows from the definition of q that $\left\lfloor \frac{M_0 - M}{N} \right\rfloor = 0$ (i.e., $M_0 < M + N$). Therefore, $S + L \leq \left\lfloor \frac{M_0}{d} \right\rfloor \leq \frac{M + N - d}{d}$. As a result, $\left(S + L_S\right)d \leq M + N - d$, hence

$$
D - D' \geq M + N - d - \left(S + L_S\right)d \geq 0.
\tag{6.47}
$$

Therefore, in this case, the system can achieve more DoF when one macro user is served.

- **Case 2:** $0 < M - Sd < N$, $M - d > N$, $N \geq Sd$

In this case, D and D' can be expressed as

$$
D = M + N - d + qN + \left(J - q\right)\left(N - d\right).
\tag{6.48}
$$

$$
\begin{aligned}
D' &= \left(S + L_S\right)d + q'\left(M - Sd\right) + \left(J - q'\right)\left(N - Sd\right) \\
&\leq \left(S + L_S\right)d + q'N + \left(J - q'\right)\left(N - Sd\right) \\
&\leq \left(S + L_S\right)d + qN + \left(J - q\right)\left(N - Sd\right).
\end{aligned}
\tag{6.49}
$$

Therefore, we have

$$
\begin{aligned}
D - D' &\geq M + N - d + \left(J - q\right)\left(N - d\right) \\
&\quad - \left(S + L_S\right)d - \left(J - q\right)\left(N - Sd\right) \\
&= \left(M + N - d - L_S d\right) - Sd \\
&\quad + \left(J - q\right)\left(S - 1\right)d \\
&\geq d - Sd + \left(J - q\right)\left(S - 1\right)d \\
&= \left(J - q - 1\right)\left(S - 1\right)d.
\end{aligned}
\tag{6.50}
$$

If $q < J$, then $D - D' \geq 0$.

- **Case 3:** $0 < M - Sd < N$, $M - d > N$, $N < Sd$

In this case, D is the same as (6.48), while D' is

$$
\begin{aligned}
D' &= \left(S + L_S\right)d + q'\left(M - Sd\right) \\
&\leq \left(S + L_S\right)d + q'N.
\end{aligned}
\tag{6.51}
$$

Note that when $q < J$, q and q' are $\left\lfloor \frac{M_0 - M}{N} \right\rfloor$ and $\left\lfloor \frac{M_0 - \left(S + L_S\right)d}{N} \right\rfloor$, respectively. Because M, N are divisible by d, M_0 can be expressed as

$$
M_0 = qN + M + ad + \varepsilon,
\tag{6.52}
$$

$$
M_0 = q'N + \left(S + L_S\right)d + a'd + \varepsilon,
\tag{6.53}
$$

where a, a' are integers, and $ad \leq N - d$, $a'd \leq N - d$, $\varepsilon < d$. Thus, we have

$$
\begin{aligned}
D - D' \geq & M + N - d + qN - (S + L_S)\,d - q'N \\
= & M_0 + N - d - ad - \varepsilon - (M_0 - a'd - \varepsilon) \\
= & N - d - (ad - a'd) \\
\geq & 0.
\end{aligned}
\tag{6.54}
$$

Therefore, if $q < J$, then $D \geq D'$.

- **Case 4:** $M - Sd \geq N$, $N > Sd$

In this case, $M - d \geq N$. Thus, D is the same as (6.48). D' is given by

$$
\begin{aligned}
D' &= (S + L_S)\,d + q'N + (J - q')\,(N - Sd) \\
&\leq (S + L_S)\,d + qN + (J - q)\,(N - Sd).
\end{aligned}
\tag{6.55}
$$

Thus, if $q < J$, we have

$$
\begin{aligned}
D - D' &\geq M + N - d - (S + L_S)\,d + (J - q)\,(S - 1)\,d \\
&\geq M + N - d - (S + L_S)\,d + (S - 1)d \\
&= M + N - d - (L_S + 1)\,d \\
&\geq 0.
\end{aligned}
\tag{6.56}
$$

- **Case 5:** $M - Sd \geq N$, $N \leq Sd$

In this case, D is the same as (6.48), while D' can be expressed as

$$
D' = (S + L_S)\,d + q'N.
\tag{6.57}
$$

Note that when $q < J$, M_0 can be expressed as (6.52) and (6.53), thus D and D' can be represented as

$$
\begin{aligned}
D &= M_0 + N - d - ad - \varepsilon + (J - q)\,(N - d), \tag{6.58} \\
D' &= M_0 - a'd - \varepsilon. \tag{6.59}
\end{aligned}
$$

Therefore, we can obtain that

$$
\begin{aligned}
D - D' &\geq M_0 + N - d - ad - \varepsilon - (M_0 - a'd - \varepsilon) \\
&= N - d - (ad - a'd) \\
&\geq 0.
\end{aligned}
\tag{6.60}
$$

From the above analysis we can see that the system can achieve more DoF by serving one macro user than by serving $S\,(2 \leq S < M/d)$ macro users under all cases if $q < J$.

$\qquad\qquad\qquad\qquad\qquad\qquad\qquad\qquad\qquad\qquad\qquad\qquad\qquad\qquad\qquad\square$

Theorem 1 tells us that under the situation that $q < J$, $S < M/d$, the system can achieve more DoF when one macro user is served. However, for other situations, it is difficult to determine how many macro users should be

---

Algorithm 6.1 Algorithm for finding the optimal value of S

1: **Input:** M_0, M, N, d, J
2: **Output:** $S^\star$, $L_S^\star$
3: Compute D according to (6.23);
4: **if** $q < J$ and $D \geq \lfloor M_0/d \rfloor d$ **then**
5:     $S^\star \leftarrow 1$; $L_S^\star \leftarrow (M + N)/d - 2$;
6: **else**
7:     $D_{\max} \leftarrow D$;
8:     **if** $q = J$ **then**
9:       **for** $S \leftarrow 2$ **to** $M/d - 1$ **do**
10:         Compute L_S according to (6.39);
11:         Compute D' according to (6.34);
12:         **if** $D' > D_{\max}$ **then**
13:           $D_{\max} \leftarrow D'$; $S^\star \leftarrow S$; $L_S^\star \leftarrow L_S$;
14:         **end if**
15:       **end for**
16:     **end if**
17:     **if** $D_{\max} < \lfloor M_0/d \rfloor d$ **then**
18:       $S^\star \leftarrow \left\lceil \Upsilon - \frac{\Upsilon(N-d)}{d\Upsilon - M} \right\rceil$; $L_S^\star \leftarrow \lfloor M_0/d \rfloor - S^\star$;
19:     **end if**
20: **end if**

---

served. Therefore, we designed an algorithm to find the optimal value of S so as to maximize the system DoF for a given system configuration. In the following, we briefly describe the main idea of the algorithm. The details of the algorithm are given by Algorithm 6.1.

As shown in Theorem 1, when $q < J$ and S is in the range $[1, M/d)$, the optimal S is one, and the maximum achievable DoF is D. When $q < J$ and $S \geq M/d$, links in subsystem 2 cannot transmit because they do not have interference-free signal space. Therefore, in this situation, only links in subsystem 1 transmit and IA is achieved by performing GHIA scheme. From (6.36), we can see that the achievable DoF of GHIA is upper bounded by $\lfloor M_0/d \rfloor d$. Therefore, if $q < J$ and $D \geq \lfloor M_0/d \rfloor d$, the maximum achievable DoF is D and the optimal S is one (line 3–line 5). Otherwise, if $q < J$ and $D < \lfloor M_0/d \rfloor d$, the maximum achievable DoF is $\lfloor M_0/d \rfloor d$ and the optimal $S \geq M/d$. From (6.39), we can see that when GHIA achieves $\lfloor M_0/d \rfloor d$ DoF, the following condition must be satisfied

$$\frac{\Phi + \sqrt{\Phi^2 + 4Sd(N-d)}}{2d} \geq \left\lfloor \frac{M_0}{d} \right\rfloor - S. \tag{6.61}$$

According to (6.61), we can obtain that

$$S \geq \left\lceil \Upsilon - \frac{\Upsilon(N-d)}{d\Upsilon - M} \right\rceil, \tag{6.62}$$

where $\Upsilon = \lfloor \frac{M_0}{d} \rfloor$. Note that there many values of S exist that could make the system achieve $\lfloor M_0/d \rfloor d$ DoF. Here, we choose S as $\left\lceil \Upsilon - \frac{\Upsilon(N-d)}{d\Upsilon-M} \right\rceil$ so that more picocells can be enabled to transmit simultaneously (line 17–line 19).

According to the definition of q ($q = \min\left\{\lfloor \frac{M_0-M}{N} \rfloor, J\right\}$) q may also equal to J. For the situation that $q = J$, when $S \geq M/d$, only links in subsystem 1 can transmit and the maximum achievable DoF is $\lfloor M_0/d \rfloor d$. However, when $1 \leq S < M/d$, it is difficult to find the optimal S. Therefore, we search for the S that achieves the maximum DoF in the range $1 \leq S < M/d$ (line 7–line 16) and let $D_{\max} = \max_{1 \leq S < M/d} D'$, $S^\star = \arg\max_{1 \leq S < M/d} D'$. Then, we compare $D_{\max}$ with $\lfloor M_0/d \rfloor d$. If $D_{\max} < \lfloor M_0/d \rfloor d$, the optimal S is $\left\lceil \Upsilon - \frac{\Upsilon(N-d)}{d\Upsilon-M} \right\rceil$ (line 17–line 19). Otherwise, the optimal S is $S^\star$.

6.2.3 Extending to General Scenarios

In the previous section, in order to completely cancel the cross-tier interference, we treat all interference from pico BSs to macro users as strong interference and align this interference in interference subspaces of the macro users. However, when the pico BSs are far away from the macro users, they will cause negligible interference to the macro users. Therefore, in this section, we consider a more general scenario where some pico BSs in subsystem 2 are also disconnected with the macro users. Meanwhile, we extend the proposed two-stage IA scheme to this scenario.

6.2.3.1 One Macro User

When one user is served by the macro BS, we divide the system into three subsystems as shown in Figure 6.4. The links in subsystem 1 form a fully connected interference network. Each pico BS in subsystem 2a is connected with the macro user and its served user, while each pico BS in subsystem 2b is only connected with its served user. For easy representation, we denote the links in subsystem 2a and 2b by $\mathcal{J}_a$ and $\mathcal{J}_b$, respectively. Note that the macro BS could use its redundant antennas to cancel its interference to some links in subsystem 2. We denote $\hat{\mathcal{J}}_a$ and $\hat{\mathcal{J}}_b$ by the links in subsystem 2a and 2b with interference from macro BS being canceled by the macro BS, respectively.

The design of precoding and decoding matrices for links in subsystem 1 and 2a are the same as in Section 6.2.2.2. The links in subsystem 2b do not need to align the transmitted signals in interference subspace of the macro user. Thus, only condition (6.6) should be satisfied. As a result, the DoF can be achieved by each link in $\mathcal{J}_b$ and $\hat{\mathcal{J}}_b$ are $\min\{M, N - d\} = N - d$ and $\min\{M, N\} = N$, respectively. Note that the DoF of each link in $\hat{\mathcal{J}}_a$ is $\min\{M - d, N\} \leq N$. Therefore, we use the redundant antennas of the macro BS to cancel interference to links in subsystem 2b first, so that more DoF can be achieved. In the following, we give the precoding and decoding matrices design for subsystem 2b.

The links in $\hat{\mathcal{J}}_b$ are interference-free. Therefore, the optimal precoding and

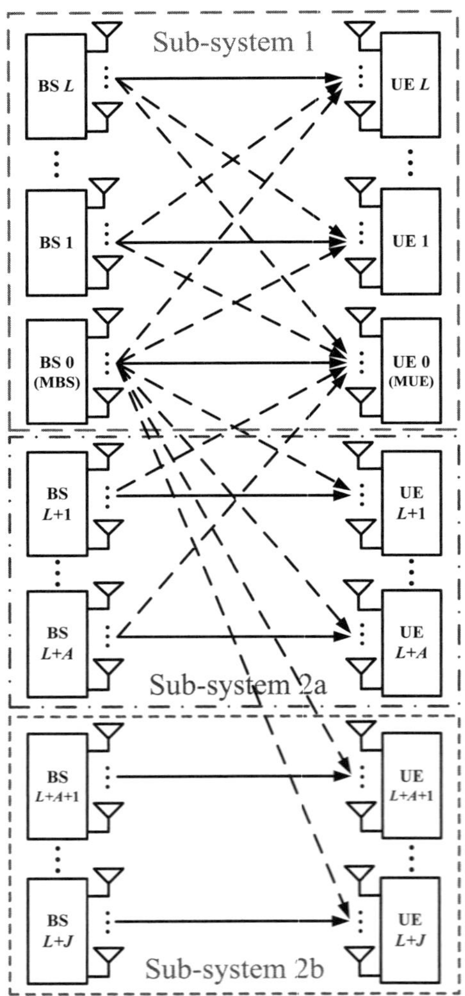

Figure 6.4: Generalized PCIN (solid arrow: desired signal, dashed arrow: interference signal).

decoding matrices of links in $\hat{\mathcal{J}}_b$ can be obtained by performing SVD on the direct channel matrices. For links in $\mathcal{J}_b \backslash \hat{\mathcal{J}}_b$, users need to cancel interference from the macro BS. We get the precoding and decoding matrices by executing SVD on the equivalent direct channel matrices

$$\widetilde{\mathbf{H}}_{j,j} = \mathbf{F}_j^\dagger \mathbf{H}_{j,j}, j \in \mathcal{J}_b \backslash \hat{\mathcal{J}}_b, \tag{6.63}$$

where $\mathbf{F}_j$ is the subspace that is not interfered by the macro BS and is the same as (6.20).

Denote the number of picocell links in subsystem 2a by A, the DoF achieved by the system can be expressed by

$$D = \underbrace{M + N - d}_{\text{the DoF achieved by subsystem1}} + \underbrace{\min\{J - A, q\} N}_{\text{the DoF achieved by links in } \hat{\mathcal{J}}_b}$$

$$+ \underbrace{(J - A - \min\{J - A, q\})(N - d)}_{\text{the DoF achieved by links in } \mathcal{J}_b \backslash \hat{\mathcal{J}}_b}$$

$$+ \underbrace{\min\{A, q - \min\{J - A, q\}\} \min\{M - d, N\}}_{\text{the DoF achieved by links in } \hat{\mathcal{J}}_a}$$

$$+ \underbrace{(A - \min\{A, q - \min\{J - A, q\}\})(N - d)}_{\text{the DoF achieved by links in } \mathcal{J}_a \backslash \hat{\mathcal{J}}_a}. \tag{6.64}$$

6.2.3.2   Multiple Macro Users

Now we consider the case where multiple users are served by the macro BS. In this case, each pico BS in subsystem 2 only needs to align the transmitted signal in interference subspaces of the macro users that are interfered by itself. The IA scheme design for this scenario is the same as in Section 6.2.2.3 except for the design of $\overline{\mathbf{W}}_0$ and $\overline{\mathbf{G}}_j$. We denote the macro users that are interfered by pico BS $j \in \mathcal{J}$ by $\mathcal{S}_j = \left\{ k_1^j, \ldots, k_{S_j}^j \right\} \subseteq \mathcal{S}$. When design $\overline{\mathbf{W}}_0$, we choose the q' links in subsystem 2 that have the smallest value of S_j. Then, $\overline{\mathbf{W}}_0$ is designed to cancel interference from the macro BS to the q' links. Because the link j with interference from the macro BS being canceled can convey $\min\{M - S_j d, N\}$ data streams, we can enable more data streams to be transmitted by choosing the q' links that have the smallest value of S_j. $\overline{\mathbf{G}}_j$ is designed as follows. If $\mathcal{S}_j = \varnothing$, then $\overline{\mathbf{G}}_j = \mathbf{I}_M$. Otherwise, $\overline{\mathbf{G}}_j$ is

$$\overline{\mathbf{G}}_j \in \text{null}\left(\left[\left(\mathbf{U}_{k_1^j}^\dagger \mathbf{H}_{k_1^j, j} \right)^\dagger, \ldots, \left(\mathbf{U}_{k_{S_j}^j}^\dagger \mathbf{H}_{k_{S_j}^j, j} \right)^\dagger \right]^\dagger \right), j \in \mathcal{J}. \tag{6.65}$$

6.2.4   Performance Results

In this section, we evaluate the performance of the proposed IA schemes. First, we investigate the performance of the two-stage IA schemes in an ideal partially connected network, where the long-term path gains between the disconnected BSs and users are set to zero while the others are set to one. Then, we evaluate the performance of the two-stage IA schemes in a practical HetNet scenario. In this scenario, the influence of the weak interference channel is studied. In the simulation, we consider such a system configuration: $M_0 = 2M$, $M = N$, $d \leq M/2$, $J \geq 2$. Under this configuration, the maximum DoF is achieved when one macro user is served according to Algorithm 6.1. In order to show the influence of the number of served macro users on the system performance, we also give simulation results when two macro users are served.

6.2.4.1 Performance in an Ideal Partially Connected Interference Network

The transmission power of the macro BS and pico BSs are denoted by P_m and P_p, respectively. We set $P_m = 2P_p$, and the variance of noise is one. In the generalized scenario (Section 6.2.3), when one user is served by the macro BS, the number of pico BSs that are disconnected with the macro user is $\lceil J/2 \rceil$. When two macro users are served, the number of pico BSs that are disconnected with both macro users, macro user 1 and macro user 2 are $J_1 = \lceil J/4 \rceil$, $J_2 = \lceil (J - J_1)/3 \rceil$ and $J_3 = \lceil (J - J_1 - J_2)/2 \rceil$, respectively. The remaining pico BSs are connected with both macro users. For notational brevity, we denote the two-stage IA proposed in Section 6.2.2 as TSIA, and TSIA (one MUE) and TSIA (two MUEs) represent the cases that one and two macro users are served, respectively. The two-stage IA developed in the general scenarios (Section 6.2.3) is denoted as GTSIA. Similar to TSIA, GTSIA (one MUE) and GTSIA (two MUEs) denote the cases that one and two macro users are served, respectively. As baseline schemes, the performance of GHIA [23] and ConvIA are given for comparison. In GHIA, two users are served by the macro BS, and the macrocell links perform GHIA with the picocell links in subsystem 1. The links in subsystem 2 keep silence and do not transmit data. In ConvIA, the macro BS serves one user and carries out IA with the picocell links in subsystem 1. Also, the links in subsystem 2 are silence. The IA algorithm that minimizes leakage interference proposed in [30] is used to calculate the precoding and decoding matrices in ConvIA. Each point in the figures is an average over 1,000 channel realizations.

Figure 6.5 shows the average sum rates versus the SNR. We can see that the two-stage IA schemes achieve higher average sum rates than GHIA and ConvIA. Meanwhile, by considering the partial connectivity between the pico BSs and the macro users, the system performance can be further improved. Specifically, GTSIA can achieve one more DoF than TSIA when one macro user is served, because in the general scenario one of the links in subsystem 2 is interference-free and can achieve M DoF. Although GTSIA cannot achieve more DoF than TSIA when two macro users are served, it can obtain a higher average sum rate. The reason is that when pico BSs are not connected with both macro users, they do not need to align their transmitted signals in the subspace of the disconnected macro user, thereby increasing the achievable rate. We can also

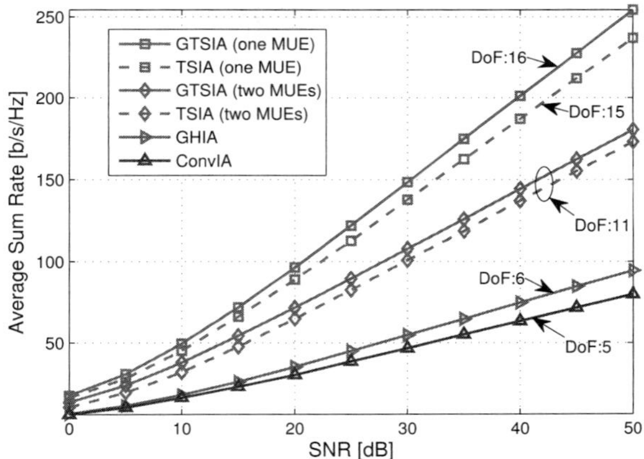

Figure 6.5: Average sum rates versus SNR under configuration $J = 5$, $M = 3$, $d = 1$.

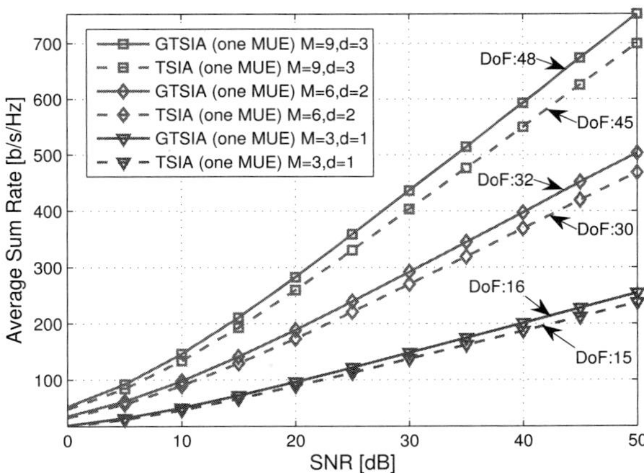

Figure 6.6: Average sum rates versus SNR under different M when $J = 5$.

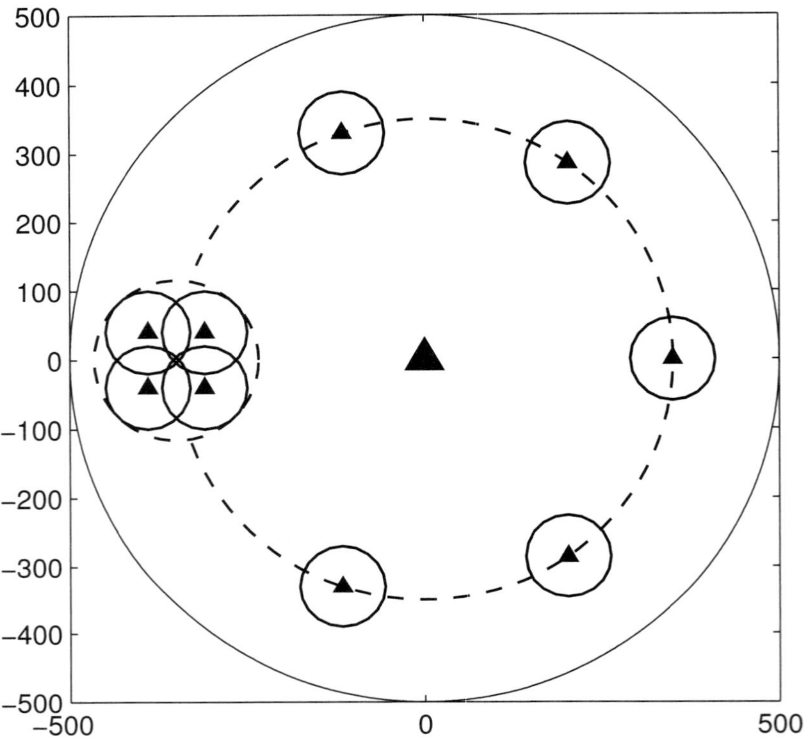

Figure 6.7: Simulation scenario.

observe from Figure 6.5 that serving two macro users reduces the system DoF. This is because links in subsystem 2 could transmit less number of data streams when two macro users are served.

Figure 6.6 shows the average sum rates versus SNR under different antenna configurations. As shown in the figure, more DoF is achieved when the number of antennas increases. We can see that under the configurations $M = 3, d = 1$, $M = 6, d = 2$, and $M = 9, d = 3$, TSIA (one MUE) can achieve 15, 30, and 45 DoF while GTSIA (two MUEs) achieves 16, 32, and 48 DoF, respectively. This coincides with the DoF given by (6.23) and (6.64).

6.2.4.2 Performance in a Practical HetNet

In this section, we evaluate the performance of the proposed two-stage IA schemes in a practical HetNet scenario, which is shown in Figure 6.7. There is a hotspot area of radius 100m in the macrocell, and four pico BSs are deployed in this area. The distance between the macro BS and the center of the hotspot is 350m. These four picocells and the macrocell form subsystem 1. The other pico BSs, which belong to subsystem 2, are deployed in an arc with a radius of 350

Table 6.1: Simulation Parameters

Parameter	Assumption
Radius of macrocell	500m
Radius of picocell	40m
Carrier frequency / bandwidth	2 GHz / 10 MHz
Number of users in each cell	10
Transmit power of macro BS (P_m)	46 dBm
Number of antennas (M)	3
DoF of each link in subsystem 1 (d)	1
Path loss from macro BS to user	$128.1+37.6\log_{10} R$[dB], R in km
Path loss from pico BS to user	$140.7+36.7\log_{10} R$[dB], R in km
Antenna pattern	0 dB (omnidirectional)
Channel model	3GPP SCM
Shadowing standard deviation	10 dB
Noise power spectral density	-174 dBm/Hz
Noise figure	9 dB

m and center at origin. The distances between neighbor pico BSs in subsystem 2 and the distances between the boundary of hotspot and the closest pico BSs in subsystem 2 are equal. The pico cells in the hotspot carry out first-stage IA with the macrocell, and the other picocells perform second-stage IA. The simulation parameters are given by Table 6.1. When one user is served by the macro BS, if $\frac{P_m g_{0,0}}{P_p g_{0,j}} \geq 30$ dB, $j \in \mathcal{J}$, then BS j is considered to be disconnected with the macro user. When the macro BS serves two users, the pico BS j that satisfies $\frac{P_m g_{k_1,0}}{P_p g_{k_1,j}} \geq 30$dB or (and) $\frac{P_m g_{k_2,0}}{P_p g_{k_2,j}} \geq 30$dB, $j \in \mathcal{J}$ is considered to be disconnected with macro user 1 or (and) macro user 2. In this scenario, we also compare to a TDMA scheme. In the TDMA scheme, data transmission is accomplished in two time slots. In the first time slot, the macro BS serves $\lfloor M_0/M \rfloor$ users and transmits M data streams to each user via multiuser MIMO techniques [33]. In the second time slot, the pico cells in subsystem 1 carry out conventional IA, and each pico BS in subsystem 2 transmits M data streams to its user. The precoding and decoding matrices of picocell links in subsystem 2 are obtained by performing SVD on the direct channel matrices. Meanwhile, the water-filling power allocation is used to improve the achievable rate.

Figure 6.8 shows the average sum rates under different transmission power of pico BSs. We can see from the figure that the two-stage IA schemes achieve higher average sum rates than other schemes. Figure 6.8 also shows that higher average sum rates are obtained when the macro BS serves one user. This is because more data streams can be conveyed by the network when one macro

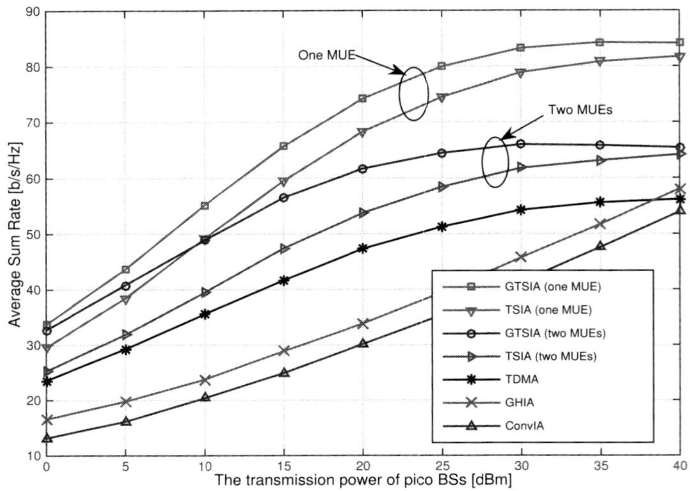

Figure 6.8: Average sum rates versus the transmission power of pico BSs under configuration $J = 5$, $M = 3$, $d = 1$.

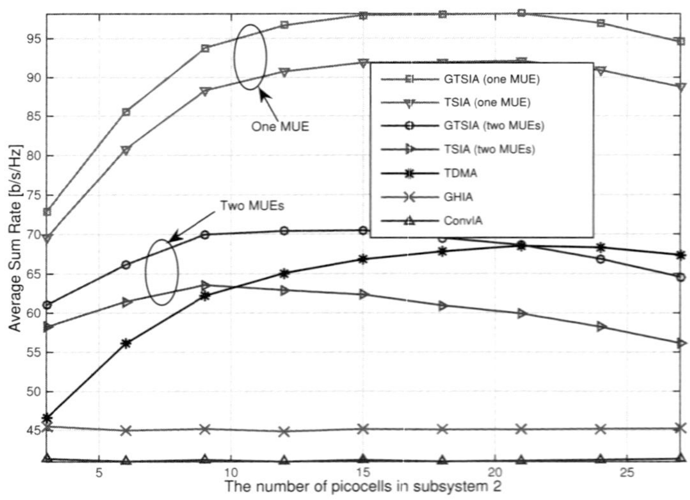

Figure 6.9: Average sum rates versus J when P_p=30 dBm.

user is served. Meanwhile, GTSIA can achieve better performance than TSIA due to the consideration of partial connectivity between pico BSs and macro users. Another phenomenon we can observe from Figure 6.8 is that the increase rates of average sum rates achieved by two-stage IA schemes decrease with the increasing of the transmission power of pico BSs. This is because the increase rate of each data stream's SINR decreases when the transmission power of pico BSs increases. In particular, when the transmission power of pico BSs is low, noise plays an important role in the SINR of each data stream. The increase of transmission power will significantly increase the SINR. However, when the transmission power is high, noise has a minor effect on SINR, and SINR is mainly determined by SIR. Meanwhile, the strength of the desired signal and interference signals increases with the same speed when the transmission power of pico BSs increases. As a result, the rate of each data stream increases slowly when transmission power is high. The TDMA scheme also has the same phenomenon for the same reason. On the contrary, the average sum rates of GHIA and ConvIA increase with constant rates because all interference among current transmitting links is canceled. However, lower average sum rates are achieved because the number of current transmitting picocells is small. In HetNets, the transmission power of pico BSs usually is lower than 30 dBm. We can see from the figure that the two-stage IA schemes can achieve a more superior performance than other schemes.

Figure 6.9 shows the average sum rates versus the number of picocells in subsystem 2. As shown in the figure, the average sum rates achieved by the two-stage IA schemes grow first and then decrease with the increasing of the number of picocells in subsystem 2. This is because when the number of picocells is not large, the cotier interference among picocells is not severe, and then the average sum rates can be improved by introducing more pico cells. However, when the number of picocells is large, the cotier interference becomes severe. As a result, the rate loss caused by the added picocells will exceed the rate increase brought on by them. Therefore, the average sum rates reduce when the number of picocells is large. A similar result can also be observed in TDMA. The GHIA and ConvIA schemes achieve constant average sum rates because only links in subsystem 1 transmit. As also shown in the figure, serving one macro user achieves higher average sum rates than serving two macro users. Moreover, by considering the partial connectivity between the pico BSs and macro users, GTSIA achieves a higher average sum rate than TSIA.

6.3 Interference Alignment in Uplink Heterogeneous Networks

6.3.1 System Model

We focus on the uplink of the a two-tier HetNet, which is shown in Figure 6.10. This HetNet consists of one macrocell and N picocells. The macrocell and picocells are overlapped with each other and use the same frequency re-

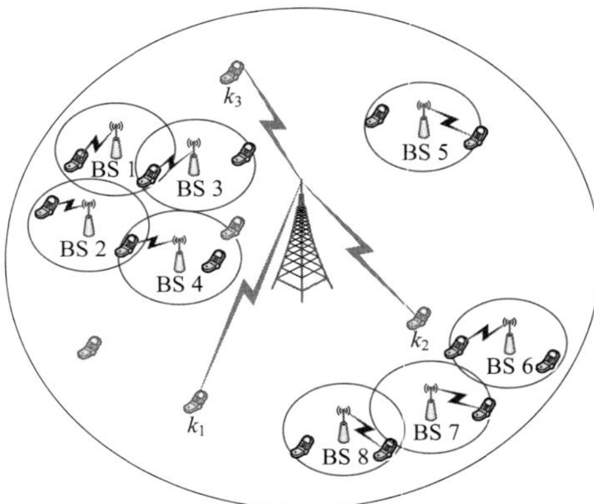

Figure 6.10: The considered HetNet.

source. For presentation convenience, we denote the BSs in the HetNet by $\{1, 2, \ldots, N, N+1\}$, where $N+1$ is the index of the macro BS. We assume that the macro BS has M_0 antennas and serves a set of users $\mathcal{K} = \{k_1, \ldots, k_L\}$ by employing multiuser MIMO, while each pico BS is equipped with M antennas and serves one user. Each user is equipped with M antennas. The channels are assumed to be frequency-flat blocking fading channels (i.e., the channel coefficients are fixed during one transmission block and change independently between blocks). We assume that the BSs are connected by wired backhaul links, and sequential interference cancellation (SIC) technique is used when BSs decode the received data. In SIC, BSs take turns to decode their desired signals. When a BS has decoded its desired signals, it sends the decoded data to the BSs that are interfered with by itself, so that these BSs can reconstruct and cancel interference signal. Usually, the macro users are far away from the serving BS compared with the pico users. Thus, in order to protect the performance of the macro users, we let the macro BS be the last one to decode data, and let the picocells that interfere with the macro users send the decoded data to the macro BS so as to cancel interference from pico users to macro users. Meanwhile, we also assume that the macro users are unaware of the existence of the pico users. Thus, the pico BSs are responsible for canceling interference from the macro users. For notational brevity, the user that is served by pico BS i is also denoted by i. Thus, the received signals at pico BS i can be represented as

$$\boldsymbol{y}_i = \underbrace{\sqrt{P_i g_{ii}}\mathbf{H}_{ii}\mathbf{V}_i\boldsymbol{x}_i}_{\text{desired signals}} + \underbrace{\sum_{j=1,j\neq i}^{N} \sqrt{P_j g_{ji}}\mathbf{H}_{ji}\mathbf{V}_j\boldsymbol{x}_j}_{\text{interference from pico users}} + \underbrace{\sum_{k_l \in \mathcal{K}} \sqrt{P_{k_l} g_{k_l i}}\mathbf{H}_{k_l i}\mathbf{V}_{k_l}\boldsymbol{x}_{k_l}}_{\text{interference from macro users}} + \boldsymbol{n}_i,$$

$$(6.66)$$

where $\boldsymbol{x}_i$, $\mathbf{V}_i$, and P_i are the the transmitted data, the precoding matrix, and the transmission power of user i, respectively. g_{ji} and $\mathbf{H}_{ji} \in \mathbb{C}^{M \times M}$ are the long-term path gain and channel matrix from user j to pico BS i, respectively, and $\boldsymbol{n}_i$ is the additive white Gaussian noise with zero mean and variance $\sigma^2 \mathbf{I}_M$. On the right-hand side of (6.66), the first term is the desired signals of pico BS i, and the second and the third terms are interference from other pico users and the macro users, respectively. If the receiver is the macro BS, then the third term is the desired signals while the first and the second terms are interference signals.

6.3.2 IAC Scheme

In this section, we detail the proposed IAC scheme. First, we represent interference relationship among the transmission links in the network as a directed conflict graph. Then, based on the constructed directed conflict graph, we present the IAC scheme, which involves a set of constraints that should be satisfied by users, BSs, and SIC. Meanwhile, by giving the design method of the precoding and decoding vectors for each data stream, we prove the feasibility of the IAC scheme.

6.3.2.1 Construct Directed Conflict Graph

In order to clearly describe interference relationship among the transmission links in the network, we represent interference relationship among the transmission links as a directed conflict graph. We denote the directed conflict graph by $G = (V, E)$, where V and E are the vertices and the edges of G, respectively. Each vertex in G corresponding to one transmission link in the network. One edge, such as (i, j), is an ordered pair of two different vertices, and represents that there is a directed edge from vertex i to vertex j. For notational brevity, the transmission link of each picocell is denoted by the same index of the picocell, and the transmission links of the macrocell are denoted by the same indices of the macro users. If transmission link i causes interference to transmission link j, then there is an edge from vertex i to j. Let e_{ij} indicate whether exists an edge from vertex i to vertex j. If there is a directed edge from vertex i to j, then $e_{ij} = 1$; Otherwise, $e_{ij} = 0$. The value of e_{ij} is determined by

$$e_{ij} = \begin{cases} 1, & \gamma_{ij} < \hat{\gamma}, i \neq j \\ 0, & \text{otherwise} \end{cases} \tag{6.67}$$

where $i, j \in \{1, \ldots, N, k_1, \ldots, k_L\}^2$, $\hat{\gamma}$ is a threshold, and γ_{ij} is defined as $\gamma_{ij} = \frac{P_j g_{jj}}{P_i g_{ij}}$. As an illustration, we give the directed conflict graph corresponding to the HetNet shown in Figure 6.10 in Figure 6.11.

---

[2]The interuser interference among macro users can be canceled by using multiuser MIMO, thus we regard that there is no interference among macro users (i.e., $e_{ij} = 0$ $(i, j \in \{k_1, \ldots, k_L\})$).

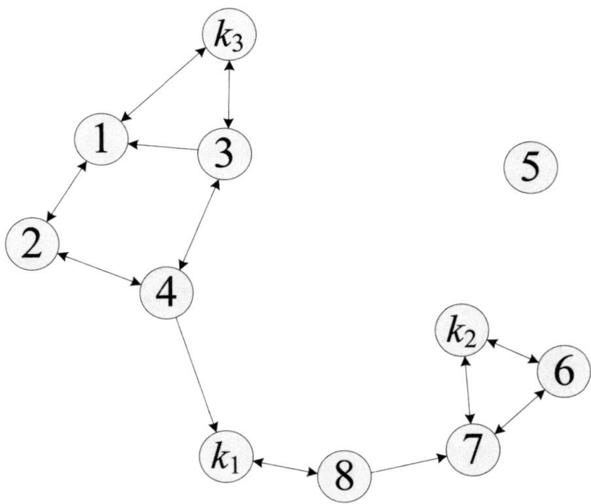

Figure 6.11: Directed conflict graph corresponding to the HetNet in Figure 6.10.

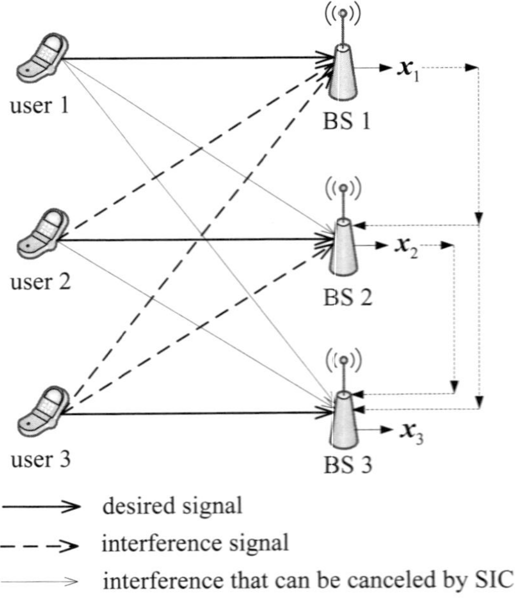

Figure 6.12: Illustration of SIC.

6.3.2.2 Design of IAC Scheme

Denote by $\mathcal{T}_j \subseteq \{1, \ldots, N, k_1, \ldots, k_L\}$ the set of users that interfere with BS j, and denote by $\mathcal{I}_i \subseteq \{1, \ldots, N, N+1\}$ the set of BSs that are interfered with by user i. Note that SIC technique is employed when BSs decode the received signals, thus the cell that has performed decoding operation will not interfere with the cells that have not decoded the data if it sends the decoded data to these cells. As shown in Figure 6.12, the decoding order of BSs is from BS 1 to BS 3 and the decoded data is sent to the BSs that have not performed data decoding, thus BS 2 will not interfered with by user 1, and BS 3 will not interfered with by user 1 and user 2. However, for the users in $\mathcal{T}_j$, if their serving BSs decode later than BS j, then they must align the transmitted signals at BS j so that BS j could decode its desired signals successfully. In Figure 6.12, user 2 and user 3 should align the transmitted signals at BS 1 so that BS 1 can decode its desired signal successfully. We use a binary variables x_{ij} to describe the SIC order between BS i and BS j. If BS i sends the decoded data to BS j for interference cancellation, then $x_{ij} = 1$. Otherwise, $x_{ij} = 0$. When all $x_{ij}\ (i, j \in \{1, \ldots, N\})$ are given, we can determine the SIC order of the whole network.

Denote by $\mathcal{S}_i\ (|\mathcal{S}_i| = \sigma_i)$ the set of data streams transmitted by user i. Suppose that cell i interferes with cell $j \in \mathcal{I}_i$ and $x_{ij} = 0$. Thus, for BS j, all date streams in $\mathcal{S}_i$ are interfering streams. We hope that as many streams from user i can be aligned at BS j as possible. Denote by $\mathcal{A}_{ij}\ (\mathcal{A}_{ij} \in \mathcal{S}_i, |\mathcal{A}_{ij}| = \alpha_{ij})$ the set of interfering data steams that can be aligned to a certain predefined interference subspace at BS j. In $\mathcal{S}_i$, there may exist a set of data streams that cannot be aligned to the predefined interference subspace of any BS in $\mathcal{I}_i$, we denote this set of data streams by $\mathcal{B}_i\ (\mathcal{B}_i \in \mathcal{S}_i, |\mathcal{B}_i| = \beta_i)$. Thus, we can obtain that

$$\mathcal{B}_i = \mathcal{S}_i \setminus \left(\cup_{j \in \mathcal{I}_i, x_{ij}=0} \mathcal{A}_{ij} \right), \forall i \in \{1, 2, \ldots, N\}. \tag{6.68}$$

In the following, we give the IAC scheme, which includes a set of constraints for users, BSs, and SIC, as well as the precoding and decoding design method for each data stream. First, for a given SIC order $x_{ij}\ (i, j \in \{1, \ldots, N\})^3$, we develop the constraints for users and BSs so as to determined how data streams should be aligned. Then, the constraints imposed by SIC are developed. Afterward, we give the precoding and decoding design method and prove the feasibility of the IAC scheme.

Constraints at users. For BS $j \in \mathcal{I}_i$, if BS i sends decoded data to BS j for interference cancellation (i.e., $x_{ij} = 1$), then user i can be regarded as causing no interference to BS j and does not need to align its data streams at BS j. For example, in Figure 6.12, the data streams transmitted by user 1 can be regarded as causing no interference to BS 2 and BS 3, thus does not need to align the transmitted signal at them. However, if $x_{ij} = 0$, the data streams transmitted by user i are interfering data streams to BS j. For each data stream in $\mathcal{S}_i$,

---

[3] Note that we let the macro BS decode last, and let the pico BSs send the decoded data to the macro BS for interference cancellation. Therefore, in the following, we mainly consider how IAC is performed among pico cells.

we align it at one BS in $\mathcal{I}'_i = \{j | j \in \mathcal{I}_i, x_{ij} = 0\}$ at most by constructing the precoding vector. Therefore, we have $\mathcal{A}_{ij_1} \cap \mathcal{A}_{ij_2} = \varnothing\, (\forall j_1, j_2 \in \mathcal{I}'_i, j_1 \neq j_2)$. Based on the above discussion, we know that when $x_{ij} = 1$, then α_{ij} should be zero, while when $x_{ij} = 0$, then $\alpha_{ij} \geq 0$. Thus, the following constraint must hold

$$x_{ij}\alpha_{ij} = 0, \forall i, j \in \{1, 2, \dots, N\}. \tag{6.69}$$

Because the data streams of user i are composed of $\mathcal{B}_i$ and $\mathcal{A}_{ij}$, the following constraint must hold at user i

$$\sum_{j \in \mathcal{I}_i} (1 - x_{ij})\,\alpha_{ij} + \beta_i = \sigma_i, \forall i \in \{1, 2, \dots, N\}. \tag{6.70}$$

Based on the directed conflict graph, (6.70) can be rewritten as

$$\sum_{j=1}^{N} e_{ij} (1 - x_{ij})\,\alpha_{ij} + \beta_i = \sigma_i, \forall i \in \{1, 2, \dots, N\}. \tag{6.71}$$

Constraints at BSs. At BS j, the interfering data streams in $\mathcal{A}_{ij}$ $(i \in \mathcal{T}_j)$ must be aligned to a predefined interference subspace. In this work, we choose the predefined interference subspace at BS j as the space spanned by data streams $\left(\cup_{k_l \in \mathcal{T}_j \cap \mathcal{K}} \mathcal{S}_{k_l}\right) \bigcup \left(\cup_{i \in \mathcal{T}_j \setminus \mathcal{K}, x_{ij}=0} \mathcal{B}_i\right)$, and let each data stream in $\mathcal{A}_{ij}$ align to one of the signal dimensions occupied by these data streams. Note that the data streams in $\mathcal{A}_{ij}$ cannot be aligned to the signal dimensions occupied by data streams in $\mathcal{B}_i$, because the these streams will be not resolvable at their intended receiver. For example, if data stream $l \in \mathcal{A}_{ij}$ (the precoding vector is $\mathbf{v}^l_i$) is aligned to data stream $l' \in \mathcal{B}_i$ (the precoding vector is $\mathbf{v}^{l'}_i$), then we have $\mathbf{H}_{ij}\mathbf{v}^l_i = \mathbf{H}_{ij}\mathbf{v}^{l'}_i$, i.e., $\mathbf{v}^l_i$ and $\mathbf{v}^{l'}_i$ are linearly dependent and are not resolvable at their intended receiver. Meanwhile, any two data streams in $\mathcal{A}_{ij}$ cannot be aligned to the same signal dimension at BS j due the same reason. Therefore, the interfering streams in $\mathcal{A}_{ij}$ should be aligned to different signal dimensions occupied by the interfering streams in $\left(\cup_{k_l \in \mathcal{T}_j \cap \mathcal{K}} \mathcal{S}_{k_l}\right) \bigcup \left(\cup_{m \in \mathcal{T}_j \setminus \mathcal{K}, x_{mj}=0}^{m \neq i} \mathcal{B}_m\right)$. Therefore, the following condition must hold if the data streams in $\mathcal{A}_{ij}$ can be aligned at BS j

$$\alpha_{ij} \leq \sum_{k_l \in \mathcal{T}_j \cap \mathcal{K}} \sigma_{k_l} + \sum_{m \in \mathcal{T}_j \setminus \mathcal{K}, m \neq i} (1 - x_{mj})\,\beta_m, \tag{6.72}$$

where $i \in \mathcal{T}_j$, $j \in \{1, 2, \dots, N\}$. By incorporating with (6.67), (6.72) can be reformulated as

$$\alpha_{ij} \leq \sum_{l=1}^{L} e_{k_l j}\sigma_{k_l} + \sum_{m=1, m \neq i}^{N} e_{mj} (1 - x_{mj})\,\beta_m. \tag{6.73}$$

If BS $i \in \mathcal{T}_j$ will send the decoded data to BS j for interference cancellation (i.e., $x_{ij} = 1$), we can regard that the interfering data streams from user i do not occupy any signal dimension of BS j. However, if $x_{ij} = 0$, the number of interfering data streams that are not aligned at BS j is $(\sigma_i - \alpha_{ij})$. Conclusively, at BS j, the number of signal dimensions occupied by the interfering pico users is $\sum_{i \in \mathcal{T}_j \backslash \mathcal{K}} (1 - x_{ij})(\sigma_i - \alpha_{ij})$. Note that the interfering streams from the macro users are not aligned and should be canceled by the pico BS. Therefore, the number of signal dimensions occupied by the macro users at BS j is $\sum_{k_l \in \mathcal{T}_j \cap \mathcal{K}} \sigma_{k_l}$[4]. Since the number of signal dimensions occupied by the interfering data streams and the desired data streams cannot exceed the total number of signal dimensions at BS j, the following constraint must hold

$$\sum_{k_l \in \mathcal{T}_j \cap \mathcal{K}} \sigma_{k_l} + \sum_{i \in \mathcal{T}_j \backslash \mathcal{K}} (1 - x_{ij})(\sigma_i - \alpha_{ij}) + \sigma_j \leq M. \tag{6.74}$$

where $j \in \{1, 2, \ldots, N\}$. Also, based on the constructed directed conflict graph, (6.74) can be reformulated as

$$\sum_{l=1}^{L} e_{k_l j} \sigma_{k_l} + \sum_{i=1}^{N} e_{ij} (1 - x_{ij})(\sigma_i - \alpha_{ij}) + \sigma_j \leq M. \tag{6.75}$$

In the above, the constraints on IA are developed for a given SIC order. Also, there some constraints exist on how to perform SIC in the network. In the following, we investigate the constraints imposed by SIC.

Constraints imposed by SIC. Note that in the signal decoding process, each cell only needs to send the decoded data to the BSs that are interfered with by itself for interference cancellation. Therefore, we have the following constraint

$$x_{ij} \leq e_{ij}. \tag{6.76}$$

Another constraint imposed by SIC is that SIC order cannot form a cycle.[5] The reason is twofold. First, if the SIC order forms a cycle, we cannot determine how SIC should be performed. As illustrated in Figure 6.13, if the SIC order forms a cycle ($x_{12} = 1$, $x_{24} = 1$, $x_{43} = 1$, $x_{31} = 1$), then we cannot determine which BS should perform data decoding first. Second, if the SIC order forms a cycle, then the BS that perform data decoding later will send decoded data to the BS that have performed data decoding, this is not causality. For the example in Figure 6.13, if we assume that BS 1 decodes data first, then BS 2, 4, and 3 will perform data decoding successively. However, BS 3 should not send its decoded data to BS 1 for interference cancellation because BS 1 has performed data decoding operation. Denote all the cycles in the constructed directed conflict graph by set $\mathcal{Q}$. The cycles in the conflict graph can be enumerated by using the efficient algorithm proposed in [34]. For each cycle with length n (i.e.,

---

[4]At the pico BS, if the number of interfering data streams from the macro users is larger than or equal to M, then the pico user can not transmit data. Therefore, we do not consider these picocells in the IAC scheme. These picocells can be eliminated from the conflict graph beforehand.

[5]A cycle is a path with the exception that the first and last vertices of the path are the same. A path is a sequence of vertices where no vertex is repeated and consecutive vertices in the sequence are adjacent vertices in the graph.

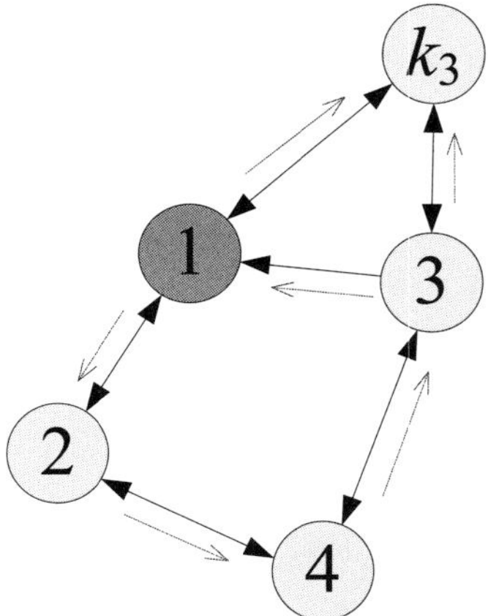

Figure 6.13: Illustration showing that SIC order cannot form a cycle.

$q = (q_1, q_2, \ldots, q_n, q_1))$, we have the following constraint

$$\sum_{i=1}^{n-1} x_{q_i q_{i+1}} + x_{q_n q_1} < n. \tag{6.77}$$

6.3.2.3 Precoding Vector Design

In this part, we show how to design the precoding vector of each data stream under the conditions that the above constraints are satisfied. For the macro users, they are not interfered with by the pico users due to the use of interference cancellation, and they do not need to align interference at the pico BS. Therefore, the existing multiuser MIMO techniques [35] can be used to design the precoding vectors of the macro users. In this work, we focus on linear transceivers. Therefore, the total number of data streams transmitted by the L macro users must satisfy the following constraint

$$\sum_{l=1}^{L} \sigma_{k_l} \leq M_0. \tag{6.78}$$

We denote the precoding vectors of the data streams of macro user k_l by $\mathcal{V}_{k_l} = \left\{ \mathbf{v}_{k_l}^1, \mathbf{v}_{k_l}^2, \ldots, \mathbf{v}_{k_l}^{\sigma_{k_l}} \right\}$, $k_l \in \mathcal{K}$.

Now, we show how to design the precoding vector of each data stream for pico user i. For the given σ_i α_{ij}, β_i, and x_{ij} that satisfy the constraints developed in Section 6.3.2.2, the precoding vectors are designed as follows. First, from the σ_i data streams, user i chooses β_i data streams to form set $\mathcal{B}_i$. For simplicity, the data streams in $\mathcal{B}_i$ are numbered as $1, \ldots, \beta_i$. Then, the left data streams form $\mathcal{A}_{ij}$ ($j \in \mathcal{I}_i'$). In the following, we show how to design the precoding vectors for data streams in $\mathcal{B}_i$.

Then, a set of linear independent complex vectors $\{\mathbf{v}_{\mathrm{ref}}^n : 1 \leq n \leq M\}$ are generated. The size of these vectors are $M \times 1$ and all entries are not zero. Then, the precoding vector of data streams in $\mathcal{B}_i$ can be designed as

$$\mathbf{v}_i^s = \mathbf{v}_{\mathrm{ref}}^s, \tag{6.79}$$

where $i \in \{1, 2, \ldots, N\}$, $s \in \{1, 2, \ldots, \beta_i\}$.

Now, we design the precoding vectors for data streams in $\mathcal{A}_{ij}$. Recall that each data stream in $\mathcal{A}_{ij}$ should be aligned to one of the signal dimensions occupied by the data streams in $\left(\cup_{k_l \in \mathcal{T}_j \cap \mathcal{K}} \mathcal{S}_{k_l}\right) \bigcup \left(\cup_{m \in \mathcal{T}_j \backslash \mathcal{K}, x_{mj}=0}^{m \neq i} \mathcal{B}_m\right)$. At BS j, the signal dimensions occupied by the macro users are $\cup_{k_l \in \mathcal{T}_j \cap \mathcal{K}} \left\{\mathbf{H}_{k_l j} \mathbf{v}_{k_l}^1, \ldots, \mathbf{H}_{k_l j} \mathbf{v}_{k_l}^{\sigma_{k_l}}\right\}$ and signal dimensions occupied by the unaligned data streams from the pico users are $\cup_{m \in \mathcal{T}_j \backslash \mathcal{K}, x_{mj}=0}^{m \neq i} \{\mathbf{H}_{mj} \mathbf{v}_m^1, \ldots, \mathbf{H}_{mj} \mathbf{v}_m^{\beta_i}\}$. Therefore, the precoding vector of each data stream in $\mathcal{A}_{ij}$ should be aligned to one of these signal dimensions at BS j; that is,

$$\mathbf{H}_{ij} \mathbf{v}_i^s = \mathbf{H}_{i'j} \mathbf{v}_{i'}^{s'}, \tag{6.80}$$

where $\mathbf{v}_i^s$ $(s \in \mathcal{A}_{ij})$ is the precoding vector of data stream s in $\mathcal{A}_{ij}$, and $\mathbf{v}_{i'}^{s'}$ is one of the precoding coding vector of data streams in $\left(\cup_{k_l \in \mathcal{T}_j \cap \mathcal{K}} \mathcal{S}_{k_l}\right) \bigcup \left(\cup_{m \in \mathcal{T}_j \backslash \mathcal{K}, x_{mj}=0}^{m \neq i} \mathcal{B}_m\right)$. Therefore, the precoding vector of data stream $s \in \mathcal{A}_{ij}$ is

$$\mathbf{v}_i^s = \mathbf{H}_{ij}^{-1} \mathbf{H}_{i'j} \mathbf{v}_{i'}^{s'}. \tag{6.81}$$

Different data streams in $\mathcal{A}_{ij}$ are aligned to different signal dimensions occupied by data streams in $\left(\cup_{k_l \in \mathcal{T}_j \cap \mathcal{K}} \mathcal{S}_{k_l}\right) \bigcup \left(\cup_{m \in \mathcal{T}_j \backslash \mathcal{K}, x_{mj}=0}^{m \neq i} \mathcal{B}_m\right)$.

In the decoding process, each BS employs a zero-forcing decoder to decode its desired data.

6.3.2.4 Feasibility of the IAC Scheme

In this part, we investigate the feasibility of the proposed IAC scheme and provide the following theorem.

Theorem 3. The IAC scheme is feasible if the developed constraints are satisfied and the proposed precoding vector design method is used.

Proof. For the macro users, the MU-MIMO technique is used to design their precoding vectors. This ensures that each data stream is resolvable at the macro BS and can be decoded. In the following, we prove that the IAC scheme is feasible for picocells.

If the IAC scheme is feasible, for pico user $j \in \{1, \ldots, N\}$ and data stream

$s \in \mathcal{S}_j$, the following conditions must hold

$$\left(\mathbf{u}_j^s\right)^\dagger \mathbf{H}_{k_l j} \mathbf{v}_{k_l}^{s'} = 0, k_l \in \mathcal{T}_j \cap \mathcal{K}, s' \in \mathcal{S}_{k_l}, \tag{6.82}$$

$$\left(\mathbf{u}_j^s\right)^\dagger \mathbf{H}_{ij} \mathbf{v}_i^{s'} = 0, i \in \mathcal{T}_j \backslash \mathcal{K}, s' \in \mathcal{S}_i, x_{ij} = 0, \tag{6.83}$$

$$\left(\mathbf{u}_j^s\right)^\dagger \mathbf{H}_{jj} \mathbf{v}_j^{s'} = 0, s' \in \mathcal{S}_j, s' \neq s, \tag{6.84}$$

$$\left(\mathbf{u}_j^s\right)^\dagger \mathbf{H}_{jj} \mathbf{v}_j^s = 1, \tag{6.85}$$

where $\mathbf{u}_i^s$ is the zero-forcing decoding vector of the desired data stream s at BS i. Condition (6.82) ensures that interference from the macro users can be completely canceled. Condition (6.83) guarantees that the interference from the other picocells can be canceled. Condition (6.84) makes sure that the inter-stream interference among the data streams transmitted by each user can be canceled, and condition (6.85) guarantees that the data streams transmitted by each user can be decoded by its serving BS.

According the feasibility conditions given by (6.82)–(6.85), in order to prove the feasibility of the IAC scheme, we need to prove two points. First, at each BS the signal subspace spanned by the desired signals is linearly independent of the signal subspace spanned by interference signals. This guarantees that conditions (6.82) and (6.83) hold. Second, the desired signals are resolvable at each BS (i.e., each desired data steam must occupy one unique signal dimension at its serving BS). This makes sure that conditions (6.84) and (6.85) hold. In the following, we first prove the first point.

At BS j, denote the set of signal dimensions occupied by the macro users by $\mathcal{R}_j^{\mathrm{MUE}}$, and denote the sets of signal dimensions occupied by the interfering streams in $\mathcal{B}_i$, $\mathcal{A}_{ij}$, and $\mathcal{S}_i \backslash \mathcal{A}_{ij}$ ($i \in \mathcal{T}_j, x_{ij} = 0$) by $\mathcal{R}_j^\beta$, $\mathcal{R}_j^\alpha$, and $\mathcal{R}_j^S$, respectively. Thus, $\mathcal{R}_j^{MUE}$, $\mathcal{R}_j^\beta$, $\mathcal{R}_j^\alpha$, and $\mathcal{R}_j^S$ can be presented as follows

$$\mathcal{R}_j^{\mathrm{MUE}} = \cup_{k_l \in \mathcal{T}_j \cap \mathcal{K}, s \in \mathcal{S}_{k_l}} \left\{ \mathbf{H}_{k_l j} \mathbf{v}_{k_l}^s \right\}, \tag{6.86}$$

$$\mathcal{R}_j^\beta = \cup_{i \in \mathcal{T}_j \backslash \mathcal{K}, x_{ij}=0, s \in \mathcal{B}_i} \left\{ \mathbf{H}_{ij} \mathbf{v}_i^s \right\}, \tag{6.87}$$

$$\mathcal{R}_j^\alpha = \cup_{i \in \mathcal{T}_j \backslash \mathcal{K}, x_{ij}=0, s \in \mathcal{A}_{ij}} \left\{ \mathbf{H}_{ij} \mathbf{v}_i^s \right\}, \tag{6.88}$$

$$\mathcal{R}_j^S = \cup_{i \in \mathcal{T}_j \backslash \mathcal{K}, x_{ij}=0, s \in \mathcal{S}_i \backslash \mathcal{A}_{ij}} \left\{ \mathbf{H}_{ij} \mathbf{v}_i^s \right\}. \tag{6.89}$$

Due to $\mathcal{B}_i \subseteq \mathcal{S}_i \backslash \mathcal{A}_{ij}$, we have $\mathcal{R}_j^\beta \subseteq \mathcal{R}_j^S$. Meanwhile, because the data streams in $\mathcal{A}_{ij}$ are aligned in α_{ij} signal dimensions occupied by data streams $\left(\cup_{k_l \in \mathcal{T}_j \cap \mathcal{K}} \mathcal{S}_{k_l}\right) \bigcup \left(\cup_{m \in \mathcal{T}_j \backslash \mathcal{K}, x_{mj}=0}^{m \neq i} \mathcal{B}_m\right)$. Therefore, we can obtain that $\mathcal{R}_j^\alpha \subseteq \mathcal{R}_j^{\mathrm{MUE}} \cup \mathcal{R}_j^\beta$. Denote by $\mathcal{R}_j^I$ the signal dimensions occupied by all interference signals at BS j. $\mathcal{R}_j^I$ can be represented as

$$\mathcal{R}_j^I = \mathcal{R}_j^{\mathrm{MUE}} \cup \left(\cup_{i \in \mathcal{T}_j \backslash \mathcal{K}, x_{ij}=0, s \in \mathcal{S}_i} \left\{ \mathbf{H}_{ij} \mathbf{v}_i^s \right\} \right). \tag{6.90}$$

Then, based on the above discussion, we can obtain the following equation.

$$\mathrm{span}\left(\mathcal{R}_j^I\right) = \mathrm{span}\left(\mathcal{R}_j^{\mathrm{MUE}} \cup \mathcal{R}_j^S \cup \mathcal{R}_j^\alpha\right) = \mathrm{span}\left(\mathcal{R}_j^{\mathrm{MUE}} \cup \mathcal{R}_j^S\right). \tag{6.91}$$

Denote by $\mathcal{R}_j^D$ the signal dimensions occupied by the desired data streams at BS j. Now, we argue that the signal subspace spanned by $\mathcal{R}_j^D$ is linearly independent of interference subspace spanned by $\mathcal{R}_j^I$. On one hand, from the constraint (6.74), we know that

$$\left| \mathcal{R}_j^I \cup \mathcal{R}_j^D \right| = \sum_{k_l \in \mathcal{T}_j \cap \mathcal{K}} \sigma_{k_l} + \sum_{i \in \mathcal{T}_j \setminus \mathcal{K}} (1 - x_{ij})(\sigma_i - \alpha_{ij}) + \sigma_j \leq M.$$

That is, the total number of signal dimensions occupied by interference signals and the desired signals cannot exceed the total available signal dimensions at BS j. On the other hand, the desired data streams of BS j are composed of two parts (i.e., the data streams in $\mathcal{B}_j$ and the data streams in $\cup_{i \in \mathcal{I}_j, x_{ji}=0} \mathcal{A}_{ji}$). At BS j, the signal dimensions occupied by the data streams in $\mathcal{B}_j$ are $\cup_{s \in \mathcal{B}_j} \left\{ \mathbf{H}_{jj} \mathbf{v}_j^s \right\}$, and the signal dimensions occupied by the data streams in $\cup_{i \in \mathcal{I}_j, x_{ji}=0} \mathcal{A}_{ji}$ are $\cup_{i \in \mathcal{I}_j, x_{ji}=0, s \in \mathcal{A}_{ji}} \left\{ \mathbf{H}_{jj} \mathbf{v}_j^s \right\}$. In this work, we assume that all the elements of the channel matrices are randomly and independently generated from continuous distributions. Therefore, due to the fact that the direct channel matrix $\mathbf{H}_{jj}$ is independent of the interfering channel matrices $\mathbf{H}_{ij}$ $(i \in \mathcal{T}_j, x_{ij} = 0)$, and the signal dimensions occupied by the interfering data streams and the desired data streams do not exceed the total number of available signal dimensions M, we can conclude that the signal dimensions occupied by the interfering data streams and the desired data streams are linearly independent of each other with probability one; that is,

$$\dim \left(\mathcal{R}_j^D \cup \mathcal{R}_j^I \right) = \dim \left(\mathcal{R}_j^D \right) + \dim \left(\mathcal{R}_j^I \right). \tag{6.92}$$

Until now, we have proved that the signal subspace spanned by $\mathcal{R}_j^D$ is linearly independent of interference subspace spanned by $\mathcal{R}_j^I$. In the following, we will prove that the desired data streams are resolvable at each BS.

According to (6.79), we know that the design of the precoding vectors of data streams in $\mathcal{B}_j$ is independent of any channel matrix. Meanwhile, according to (6.80), the precoding vectors of data streams $\cup_{i \in \mathcal{I}_j, x_{ji}=0} \mathcal{A}_{ji}$ are designed based on the interfering channel matrices $\mathbf{H}_{ji}$ $(i \in \mathcal{I}_j, x_{ji} = 0)$. Note that all the elements of the channel matrices are randomly and independently generated from continuous distributions, thus the signal dimensions occupied by data streams in $\mathcal{B}_j$ and data streams in $\cup_{i \in \mathcal{I}_j, x_{ji}=0} \mathcal{A}_{ji}$ are linearly independent of each other with probability one at BS j; that is,

$$\dim \left(\left(\cup_{s \in \mathcal{B}_j} \left\{ \mathbf{H}_{jj} \mathbf{v}_j^s \right\} \right) \cup \left(\cup_{i \in \mathcal{I}_j, x_{ji}=0, s \in \mathcal{A}_{ji}} \left\{ \mathbf{H}_{jj} \mathbf{v}_j^s \right\} \right) \right)$$
$$= \dim \left(\cup_{s \in \mathcal{B}_j} \left\{ \mathbf{H}_{jj} \mathbf{v}_j^s \right\} \right) + \dim \left(\cup_{i \in \mathcal{I}_j, x_{ji}=0, s \in \mathcal{A}_{ji}} \left\{ \mathbf{H}_{jj} \mathbf{v}_j^s \right\} \right). \tag{6.93}$$

For the data streams in $\mathcal{B}_j$, the precoding vectors are designed as (6.79). Note that the vectors in $\{ \mathbf{v}_{\text{ref}}^n : 1 \leq n \leq M \}$ are linearly independent of each other and each entry is not zero, thus based on the fact that the channel matrix $\mathbf{H}_{jj}$ has full rank, we can conclude that the signal dimensions $\cup_{s \in \mathcal{B}_j} \left\{ \mathbf{H}_{jj} \mathbf{v}_j^s \right\}$ are linearly independent of each other.

The precoding vectors of the macro users are designed based on the direct channel matrices while the precoding vectors of the data streams in $\mathcal{B}_j$ are designed independent of any channel matrix. Therefore, at BS j, the signal dimensions occupied by the data streams of the interfering macro users and the data streams in $\mathcal{B}_j$ are linearly independent of each other with probability one. According to (6.80), we know that the precoding vectors of data streams in $\mathcal{A}_{ji}$ is determined by channel matrix $\mathbf{H}_{ji}$. Note that $\{\mathbf{H}_{ji} : i \in \mathcal{I}_j, x_{ji} = 0\}$ are independent of each other. Thus, the following equation holds with probability one:

$$\dim\left(\cup_{i \in \mathcal{I}_j, x_{ji}=0, s \in \mathcal{A}_{ji}} \left\{\mathbf{H}_{jj}\mathbf{v}_j^s\right\}\right) = \sum_{i \in \mathcal{I}_j, x_{ji}=0} \dim\left(\cup_{s \in \mathcal{A}_{ji}} \left\{\mathbf{H}_{jj}\mathbf{v}_j^s\right\}\right). \quad (6.94)$$

For the data streams in $\mathcal{A}_{ji}$ (i.e., the data streams that are aligned to the same BS i), they are aligned to different signal dimensions at BS j. Because $\mathbf{H}_{jj}$ has full rank, the data steams that are aligned to the same BS are linearly independent of each other; that is,

$$\dim\left(\left\{\cup_{s \in \mathcal{A}_{ji}} \mathbf{H}_{jj}\mathbf{v}_j^s\right\}\right) = \alpha_{ji}, \forall i \in \mathcal{I}_j, x_{ji} = 0. \quad (6.95)$$

Based on (6.93), (6.94), and (6.95), we can conclude that

$$\begin{aligned}
&\dim\left(\mathcal{R}_j^D\right) \\
=&\dim\left(\cup_{s \in \mathcal{S}_j} \left\{\mathbf{H}_{jj}\mathbf{v}_j^s\right\}\right) \\
=&\dim\left(\left(\cup_{s \in \mathcal{B}_j} \left\{\mathbf{H}_{jj}\mathbf{v}_j^s\right\}\right) \cup \left(\cup_{i \in \mathcal{I}_j, x_{ji}=0, s \in \mathcal{A}_{ji}} \left\{\mathbf{H}_{jj}\mathbf{v}_j^s\right\}\right)\right) \\
=&\dim\left(\cup_{s \in \mathcal{B}_j} \left\{\mathbf{H}_{jj}\mathbf{v}_j^s\right\}\right) + \dim\left(\cup_{i \in \mathcal{I}_j, x_{ji}=0, s \in \mathcal{A}_{ji}} \left\{\mathbf{H}_{jj}\mathbf{v}_j^s\right\}\right) \\
=&\beta_j + \sum_{i \in \mathcal{I}_j, x_{ji}=0} \alpha_{ji}.
\end{aligned} \quad (6.96)$$

That is, each desired data stream of BS j occupies one signal dimension at BS j, thus can be correctly decoded.

In the above, we have proved that the signal dimensions occupied by the desired signals and the interfering signals are linearly independent of each other at each BS, and the data streams transmitted by each user are resolvable at the serving BS. Therefore, the IAC feasibility conditions (6.82)–(6.85) hold (i.e., the IAC scheme is feasible).

$\square$

6.3.3   Determining the Optimal IAC Scheme

Note that the number of data streams that can be transmitted in the whole network is determined by the SIC order and the number of data streams transmitted by each user. In this section, in order to determine the optimal IAC scheme, we construct an optimization problem based on the constraints developed in Section 6.3.2 with the aim of maximizing the number of data streams transmitted in the network. The formulation of the optimization problem is given as follows:

$$\textbf{P1:} \quad \max \ \sum_{i=1}^{N} \sigma_i$$

$$\text{s.t.} \quad \text{Constraints at users: } (6.69), (6.71);$$
$$\text{Constraints at BSs: } (6.73), (6.75);$$
$$\text{Constraints on SIC: } (6.76), (6.77).$$

The optimization problem P1 is a mixed integer nonlinear programming. We use the reformulation linearization technique [36] to eliminate the nonlinear terms in the constrains.

There are three nonlinear terms in the constraints (i.e., $x_{mj}\beta_m$, $x_{ij}\sigma_i$, and $x_{ij}\alpha_{ij}$). To eliminate these nonlinear terms, we define $\lambda_{mj} = x_{mj}\beta_m$, $\mu_{ij} = x_{ij}\sigma_i$, and $\omega_{ij} = x_{ij}\alpha_{ij}$. The replacement of $\lambda_{mj} = x_{mj}\beta_m$ requires to add the following two constraints:

$$0 \le \lambda_{mj} \le \beta_m, \tag{6.97}$$

$$\beta_m - (1 - x_{mj})M \le \lambda_{mj} \le x_{mj}M, \tag{6.98}$$

where $m, j \in \{1, 2, \ldots, N\}$.

Similarly, to replace $x_{ij}\sigma_i$ with $\mu_{ij} = x_{ij}\sigma_i$, we need to add the following constraints:

$$0 \le \mu_{ij} \le \sigma_i, \tag{6.99}$$

$$\sigma_i - (1 - x_{ij})M \le \mu_{ij} \le x_{ij}M, \tag{6.100}$$

where $i, j \in \{1, 2, \ldots, N\}$.

By letting $\omega_{ij} = x_{ij}\alpha_{ij}$, constraint (6.69) can be presented as

$$\omega_{ij} = 0. \tag{6.101}$$

where $i, j \in \{1, 2, \ldots, N\}$.

By replacing $\lambda_{mj} = x_{mj}\beta_m$, $\mu_{ij} = x_{ij}\sigma_i$, and $\omega_{ij} = x_{ij}\alpha_{ij}$ into constraints (6.71), (6.73), and (6.75), we have

$$\sum_{j=1}^{N} e_{ij}\left(\alpha_{ij} - \omega_{ij}\right) + \beta_i = \sigma_i. \tag{6.102}$$

$$\alpha_{ij} \le \sum_{l=1}^{L} e_{k_l j}\sigma_{k_l} + \sum_{m=1, m \ne i}^{N} e_{mj}\left(\beta_m - \lambda_{mj}\right). \tag{6.103}$$

$$\sum_{l=1}^{L} e_{k_l j}\sigma_{k_l} + \sum_{i=1}^{N} e_{ij}\left(\sigma_i - \alpha_{ij} - \mu_{ij} + \omega_{ij}\right) + \sigma_j \le M. \tag{6.104}$$

Consequently, the optimization problem P1 can be reformulated:

$$\textbf{P2:} \quad \max \ \sum_{i=1}^{N} \sigma_i$$

$$\text{s.t.} \quad \text{Constraints at users: } (36), (6.102);$$
$$\text{Constraints at BSs: } (6.103), (6.104);$$
$$\text{Constraints on SIC: } (6.76), (6.77)$$
$$\text{and } (32), (33), (34), (35).$$

where $\sigma_{k_l}, e_{ij}, L, N$, and M are known; x_{ij} are binary variables; $\sigma_i, \beta_i, \alpha_{ij}, \lambda_{mj}, \mu_{ij}$, and ω_{ij} are non-negative integer variables.

P2 is a mixed integer linear programming (MILP). Although the worst-case complexity of solving a general MILP problem is exponential, there are some efficient optimality/approximation algorithms, such as branch-and-bound with cutting planes to solve this kind of problem. Another approach is to use existing software to solve the MILP problems. In this work, we employ the MOSEK solver in CVX [37] to solve the MILP problem.

6.3.4   Performance Evaluation

In this section, we evaluate the performance of the IAC scheme through simulation. We give the complete results to show the average performance of the IAC scheme under different system configurations. In the simulation, we compare with an IA scheme and a TDMA scheme. In the IA scheme, the decoded data are not allowed to be exchanged among pico BSs, but the pico BSs will send the decoded data to the macro BS so as to cancel interference to the macro users. Therefore, the IA scheme can be formulated as follows.

$$\textbf{P-IA:} \quad \max \sum_{i=1}^{N} \sigma_i$$

$$\text{s.t.} \sum_{j=1}^{N} e_{ij}\alpha_{ij} + \beta_i = \sigma_i, \forall i \in \{1, 2, \ldots, N\};$$

$$\alpha_{ij} \leq \sum_{l=1}^{L} e_{k_l j}\sigma_{k_l} + \sum_{m=1, m \neq i}^{N} e_{mj}\beta_m, \forall i \in \mathcal{T}_j, j \in \{1, 2, \ldots, N\};$$

$$\sum_{l=1}^{L} e_{k_l j}\sigma_{k_l} + \sum_{i=1}^{N} e_{ij}(\sigma_i - \alpha_{ij}) + \sigma_j \leq M, \forall j \in \{1, 2, \ldots, N\}.$$

In the TDMA scheme, the data transmission is performed in two time slots. In the first time slot, the macro users performs data transmission. In the second time slot, the maximum number of picocells that do not interfere with each other are scheduled to transmit data, and each pico user transmits M data streams to its serving BS. Note that this scheme is similar to the concept of ABS used by

Table 6.2: Simulation Parameters

Parameter	Assumption
Radius of the macrocell	500 m
Radius of the picocell	40 m
Carrier frequency / bandwidth	2 GHz / 10 MHz
Path loss from macro BS to user	$128.1+37.6\log_{10} R$[dB], R in km
Path loss from pico BS to user	$140.7+36.7\log_{10} R$[dB], R in km
Antenna pattern	0 dB (omnidirectional)
Transmission power of users	20 dBm
Channel model	3GPP SCM
Noise power spectral density	-174 dBm/Hz
Noise figure	9 dB
Threshold $\widehat{\gamma}$	30 dB
M_0	6
M	4
L	6
$\sigma_{k_l}\ (l = 1, \ldots, L)$	1
Traffic model	full buffer

4G, where the macro BS keeps specific subframes empty to allow the pico BS to operate with less interference. The method of determining the picocells that transmit at the second time slot is given as follows. We let y_i denote whether picocell i is scheduled for data transmission or not. If picocell i is scheduled to transmit, then $y_i = 1$; otherwise $y_i = 0$. The picocells that are selected for data transmission at the second time slot are determined by solving the following optimization problem.

$$\textbf{P-TDMA}: \ \max \ \sum_{i=1}^{N} y_i$$

$$\text{s.t. } \delta_{ij}\left(y_i + y_j\right) \leq 1, \forall i, j \in \{1, 2, \ldots, N\}.$$

δ_{ij} is defined as

$$\delta_{ij} = \begin{cases} 1, & e_{ij} = 1 \text{ or } e_{ji} = 1; \\ 0, & \text{otherwise.} \end{cases}$$

The simulation parameters are given in Table 6.2. In the simulation, the pico-cells are uniformly and randomly deployed in the coverage area of the macrocell. The users served by each BS are also randomly dropped in the coverage area of each BS.

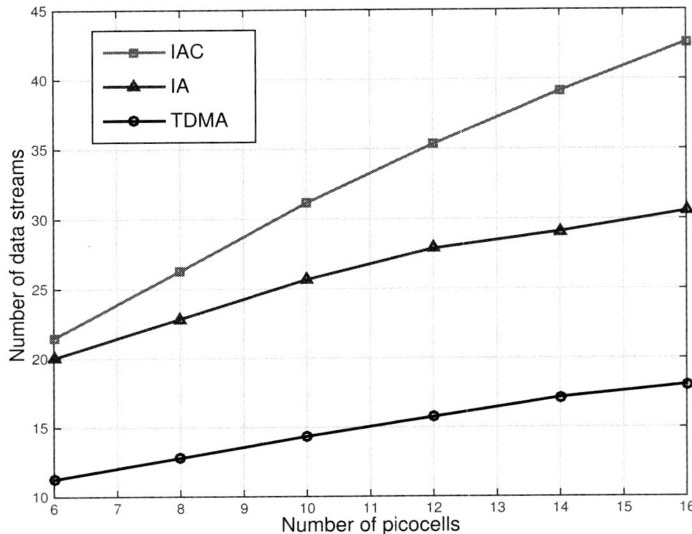

Figure 6.14: Average number of data streams versus the number of picocells.

Figures 6.14 and 6.15 show the average number of data streams and the average sum rates achieved by the network, respectively. As shown in Figure 6.14, with the increasing of the number of picocells in the network, the number of transmitted data streams increases for all schemes because more small cells can transmit simultaneously. As a result, higher sum rates are achieved by all schemes, which is shown in Figure 6.15. As also shown in Figures 6.14 and 6.15, the IAC scheme achieves the best performance, because it effectively integrates IA and interference cancellation. The TDMA scheme achieves the worst performance because it employs a resource-splitting method to avoid interference, which degrades the system DoF. We can also observe that when the number of picocells in the network is small, the IAC scheme outperforms the IA scheme slightly. This is because when the number of picocells is small, cotier interference in the network is not severe. Consequently, interference cancellation can only bring limited performance gain. With the increasing of the number of picocells in the network, more intercell interference can be canceled by interference cancellation. As a result, the IAC scheme will enable more data streams to be transmitted in the network and achieve higher sum rate.

In Figure 6.16, the average number of data streams that can be transmitted in the network under different number of antennas is given. As the figure shows, the number of transmitted data streams increases linearly with the increasing of the number of antennas for all schemes. This is because when the number of antennas increases, more signal dimensions can be used for data transmission. As a result, more data streams can be transmitted in the network. Meanwhile,

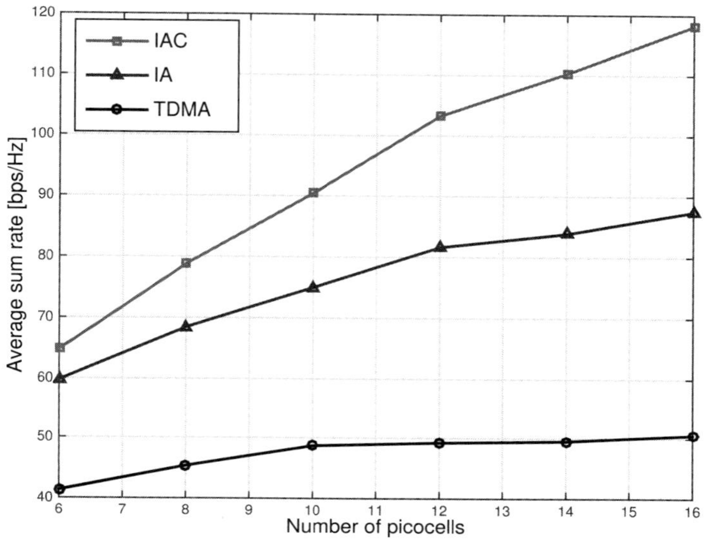

Figure 6.15: Average sum rate versus the number of pico cells.

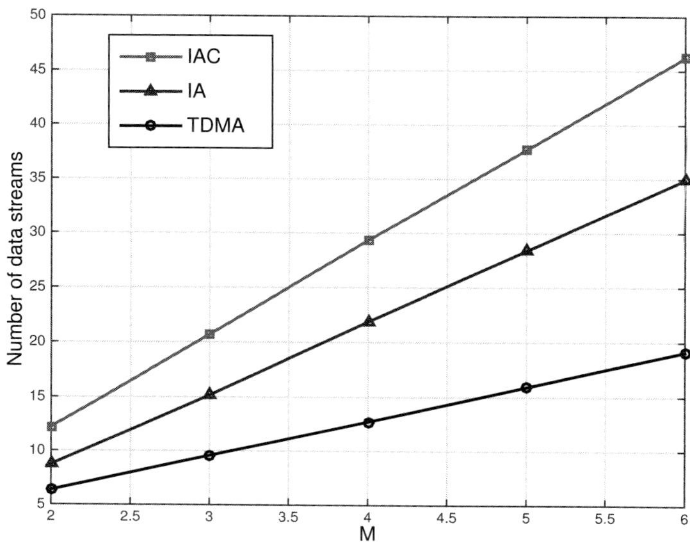

Figure 6.16: Average number of data streams versus M.

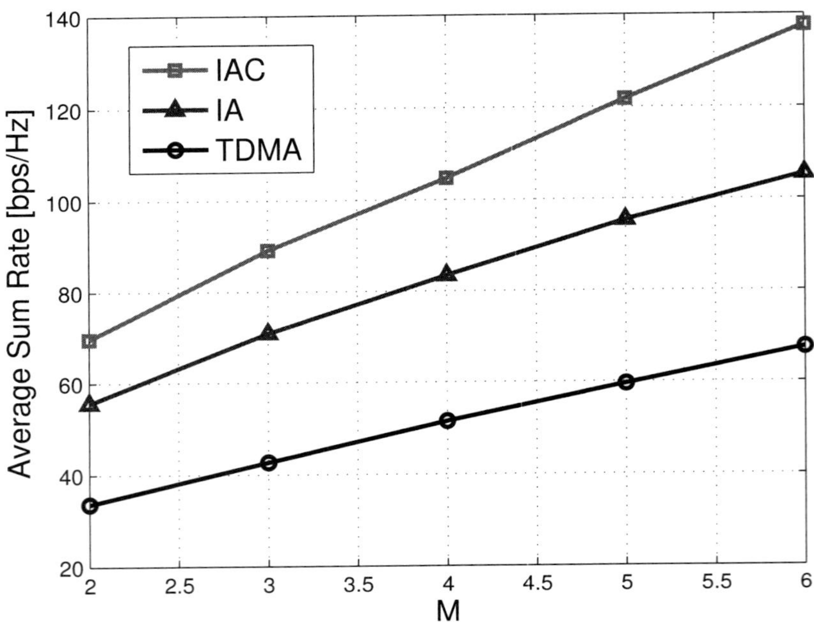

Figure 6.17: Average sum rate versus M.

as shown in Figure 6.17, the average sum rates of all schemes increase with the increasing of antenna number because more data streams can be transmitted. As also shown in Figures 6.16 and 6.17, the IAC scheme achieves the best performance due to exploiting of backhual resources.

6.4   Conclusions

In this chapter, we first proposed a two-stage IA scheme for downlink MIMO HetNets by exploiting the heterogeneity and partial connectivity structure of HetNets. In particular, two partially connected scenarios were considered. In the first scenario, we focused on partial connectivity among small cells, and proposed a two-stage IA scheme to address interference problem and increase the DoF of HetNets. Moreover, we investigated the influence of the number of served macro users on the system DoF. In particular, we derived the conditions under which serving one macro user achieves more DoF than serving multiple macro users. Meanwhile, an algorithm was designed to find the optimal number of served macro users such that the maximum system DoF can be achieved. In the second scenario, besides partial connectivity among picocells, we also considered partial connectivity between pico BSs and macro users. The two-stage IA scheme is extended to this scenario. Finally, simulation results show that the proposed two-stage IA schemes can achieve better performance than other schemes by making use of the special characteristics of HetNets. Moreover, by considering the partial connection between pico BSs and macro users, network performance can be further improved.

Then, we proposed an IAC scheme for the uplink of HetNets by taking the special characteristic of HetNets into consideration. In particular, we developed the constraints imposed by interference cancellation and developed a set of constraints at each BS and each user so as to determine how to perform IA for a given SIC order. Meanwhile, by providing the precoding vectors design method, we proved the feasibility of the IAC scheme. In order to determine the optimal SIC order and the corresponding IA scheme, we formulated the IAC scheme as an optimization problem with the aim of maximizing the number of transmitted data streams based on the developed constraints. The simulation results show that the IAC scheme can significantly improve the performance of HetNets.

REFERENCES

[1] A. Ghosh, N. Mangalvedhe, R. Ratasuk, B. Mondal, M. Cudak, E. Visotsky, T. A. Thomas, J. G. Andrews, P. Xia, H. S. Jo *et al.*, "Heterogeneous cellular networks: From theory to practice," *IEEE Commun. Mag.*, vol. 50, no. 6, pp. 54–64, Jun. 2012.

[2] V. Chandrasekhar, J. Andrews, and A. Gatherer, "Femtocell networks: A survey," *IEEE Commun. Mag.*, vol. 46, no. 9, pp. 59–67, Sep. 2008.

[3] N. Saquib, E. Hossain, L. B. Le, and D. I. Kim, "Interference management in OFDMA femtocell networks:issues and approaches," *IEEE Wireless Commun.*, vol. 19, no. 3, pp. 86–95, Jun. 2012.

[4] V. Chandrasekhar and J. Andrews, "Spectrum allocation in tiered cellular networks," *IEEE Trans. Commun.*, vol. 57, no. 10, pp. 3059–3068, Oct. 2009.

[5] Y.-S. Liang, W.-H. Chung, G.-K. Ni, I.-Y. Chen, H. Zhang, and S.-Y. Kuo, "Resource allocation with interference avoidance in OFDMA femtocell networks," *IEEE Trans. Veh. Technol.*, vol. 61, no. 5, pp. 2243–2255, Jun. 2012.

[6] M. Arslan, J. Yoon, K. Sundaresan, S. Krishnamurthy, and S. Banerjee, "A resource management system for interference mitigation in enterprise OFDMA femtocells," *IEEE/ACM Trans. Netw*, vol. 21, no. 5, pp. 1447–1460, Oct. 2013.

[7] A. Abdelnasser, E. Hossain, and D. I. Kim, "Clustering and resource allocation for dense femtocells in a two-tier cellular OFDMA network," *IEEE Trans. Wireless Commun.*, vol. 13, no. 3, pp. 1628–1641, Mar. 2014.

[8] D. Lee, H. Seo, B. Clerckx, E. Hardouin, D. Mazzarese, S. Nagata, and K. Sayana, "Coordinated multipoint transmission and reception in LTE-advanced: Deployment scenarios and operational challenges," *IEEE Commun. Mag.*, vol. 50, no. 2, pp. 148–155, February 2012.

[9] D. Gesbert, S. Hanly, H. Huang, S. Shamai Shitz, O. Simeone, and W. Yu, "Multi-cell MIMO cooperative networks: A new look at interference," *IEEE J. Select. Areas Commun.*, vol. 28, no. 9, pp. 1380–1408, Dec. 2010.

[10] G. Liu, M. Sheng, X. Wang, Y. Li, and J. Li, "Joint interference alignment and avoidance for downlink heterogeneous networks," *IEEE Commun. Lett.*, vol. 18, no. 8, pp. 1431–1434, Aug. 2014.

[11] V. R. Cadambe and S. A. Jafar, "Interference alignment and spatial degrees of freedom for the K user interference channel," *IEEE Trans. Inf. Theory*, vol. 54, pp. 3425–3441, Aug. 2008.

[12] C. Suh, M. Ho, and D. Tse, "Downlink interference alignment," *IEEE Trans. Commun.*, vol. 59, no. 9, pp. 2616–2626, Sep. 2011.

[13] R. Tresch and M. Guillaud, "Clustered interference alignment in large cellular networks," in *Proc. IEEE PIMRC*, Tokyo, Sept 2009, pp. 1024–1028.

[14] W. Shin, N. Lee, J.-B. Lim, C. Shin, and K. Jang, "On the design of interference alignment scheme for two-cell MIMO interfering broadcast channels," *IEEE Trans. Wireless Commun.*, vol. 10, no. 2, pp. 437–442, Feb. 2011.

[15] J. Tang and S. Lambotharan, "Interference alignment techniques for MIMO multi-cell interfering broadcast channels," *IEEE Trans. Commun.*, vol. 61, no. 1, pp. 164–175, Jan. 2013.

[16] C. M. Yetis, T. Gou, S. A. Jafar, and A. H. Kayran, "On feasibility of interference alignment in MIMO interference networks," *IEEE Trans. Signal Process.*, vol. 58, no. 9, pp. 4771–4782, Sep. 2010.

[17] M. Razaviyayn, G. Lyubeznik, and Z.-Q. Luo, "On the degrees of freedom achievable through interference alignment in a MIMO interference channel," *IEEE Trans. Signal Process.*, vol. 60, no. 2, pp. 812–821, Feb. 2012.

[18] L. Ruan, V. Lau, and M. Win, "The feasibility conditions for interference alignment in MIMO networks," *IEEE Trans. Signal Process.*, vol. 61, no. 8, pp. 2066–2077, Apr. 2013.

[19] B. Guler and A. Yener, "Selective interference alignment for MIMO cognitive femtocell networks," *IEEE J. Select. Areas Commun.*, vol. 32, no. 3, pp. 439–450, Mar. 2014.

[20] S. Sharma, S. Chatzinotas, and B. Ottersten, "Interference alignment for spectral coexistence of heterogeneous networks," *EURASIP Journal on Wireless Communications and Networking*, vol. 2013, no. 1, pp. 1–14, Feb. 2013.

[21] F. Pantisano, M. Bennis, W. Saad, M. Debbah, and M. Latva-aho, "Interference alignment for cooperative femtocell networks: A game-theoretic approach," *IEEE Trans. Mob. Comput.*, vol. 12, no. 99, pp. 2233–2246, Nov. 2013.

[22] S. Ben Halima and A. Saadani, "Joint clustering and interference alignment for overloaded femtocell networks," in *Proc. IEEE WCNC*, Shanghai, Apr. 2012, pp. 1229–1233.

[23] W. Shin, W. Noh, K. Jang, and H.-H. Choi, "Hierarchical interference alignment for downlink heterogeneous networks," *IEEE Trans. Wireless Commun.*, vol. 11, no. 12, pp. 4549–4559, Dec. 2012.

[24] M. Guillaud and D. Gesbert, "Interference alignment in the partially connected K-user MIMO interference channel," in *Proc. EUSIPCO*, Barcelona, Spain, Aug. 2011, pp. 1–5.

[25] M. Khatiwada and S. W. Choi, "On the interference management for K-user partially connected fading interference channels," *IEEE Trans. Commun.*, vol. 60, no. 12, pp. 3717–3725, Dec. 2012.

[26] L. Ruan, V. Lau, and X. Rao, "Interference alignment for partially connected MIMO cellular networks," *IEEE Trans. Signal Process.*, vol. 60, no. 7, pp. 3692–3701, Jul. 2012.

[27] S. Gollakota, S. D. Perli, and D. Katabi, "Interference alignment and cancellation," in *Proc. ACM SIGCOMM*, New York, USA, 2009, pp. 159–170.

[28] X. Qu and C. Kang, "A closed-form solution to implement interference alignment and cancellation for a Gaussian interference multiple access channel," *IEEE Wireless Commun.*, vol. 13, no. 2, pp. 710–723, Feb. 2014.

[29] V. Ntranos, M. Maddah-Ali, and G. Caire, "Cellular interference alignment," in *Proc. IEEE ISIT*, Jun. 2014, pp. 1598–1602.

[30] K. Gomadam, V. Cadambe, and S. Jafar, "A distributed numerical approach to interference alignment and applications to wireless interference networks," *IEEE Trans. Inf. Theory*, vol. 57, no. 6, pp. 3309–3322, Jun. 2011.

[31] R. Tresch, M. Guillaud, and E. Riegler, "On the achievability of interference alignment in the K-user constant MIMO interference channel," in *Proc. IEEE/SP SSP*, Cardiff, UK, 2009, pp. 277–280.

[32] S. Peters and R. Heath, "Interference alignment via alternating minimization," in *Proc. IEEE International Conference on Acoustics, Speech and Signal Processing (ICASSP)*, Taipei, April 2009, pp. 2445–2448.

[33] D. Tse, *Fundamentals of wireless communication.* Cambridge university press, 2005.

[34] J. C. Tiernan, "An efficient search algorithm to find the elementary circuits of a graph," *Commun. ACM*, vol. 13, no. 12, pp. 722–726, Dec. 1970.

[35] D. Gesbert, M. Kountouris, R. Heath, C.-B. Chae, and T. Salzer, "Shifting the MIMO paradigm," *IEEE Signal Processing Mag.*, vol. 24, no. 5, pp. 36–46, Sept. 2007.

[36] H. D. Sherali and W. P. Adams, *A reformulation-linearization technique for solving discrete and continuous nonconvex problems.* Springer Science & Business Media, 1998, vol. 31.

[37] I. CVX Research, "CVX: Matlab software for disciplined convex programming, version 2.0," http://cvxr.com/cvx, Aug. 2012.

About the Authors

Jiandong Li is a vice president and professor at Xidian University, Xi'an, China. He received his B.S., M.S., and Ph.D. in communication and electronic systems from Xidian University, Xi'an, China.

Min Sheng is a professor with the State Key Laboratory of Integrated Service Networks at Xidian University, Xi'an, China. She received her M.S. and Ph.D. in communication and information systems from Xidian University, Xi'an, China.

Xijun Wang is an associate professor with the School of Telecommunications Engineering at Xidian University, Xi'an, China. He received his B.S. in communications engineering from Xidian University, Xi'an, China and his Ph.D. in information and communication engineering from Tsinghua University, Beijing, China.

Hongguang Sun is a lecturer with the School of Telecommunications Engineering at Xidian University, Xi'an, China. He received his B.S. in electronic and information engineering from Northeastern University, Qinhuangdao, China and his Ph.D. in communication and information systems from Xidian University, Xi'an, China.

Index

GPRS: Gateway to Third Generation Mobile Networks,
Gunnar Heine and Holger Sagkob

GSM and Personal Communications Handbook,
Siegmund M. Redl, Matthias K. Weber, and
Malcolm W. Oliphant

GSM Networks: Protocols, Terminology, and Implementation,
Gunnar Heine

GSM System Engineering, Asha Mehrotra

Handbook of Land-Mobile Radio System Coverage, Garry C. Hess

Handbook of Mobile Radio Networks, Sami Tabbane

High-Speed Wireless ATM and LANs, Benny Bing

Inside Bluetooth Low Energy, Second Edition, Naresh Gupta

Interference Analysis and Reduction for Wireless Systems,
Peter Stavroulakis

*Interference and Resource Management in Heterogeneous
Wireless Networks,* Jiandong Li, Min Sheng, Xijun Wang, and
Hongguang Sun

Internet Technologies for Fixed and Mobile Networks,
Toni Janevski

Introduction to 3G Mobile Communications, Second Edition,
Juha Korhonen

Introduction to 4G Mobile Communications, Juha Korhonen

Introduction to Communication Systems Simulation,
Maurice Schiff

Introduction to Digital Professional Mobile Radio,
Hans-Peter A. Ketterling

Introduction to GPS: The Global Positioning System,
Ahmed El-Rabbany

An Introduction to GSM, Siegmund M. Redl, Matthias K. Weber,
and Malcolm W. Oliphant

Wireless Communications in Developing Countries: Cellular and Satellite Systems, Rachael E. Schwartz

Wireless Communications Evolution to 3G and Beyond, Saad Z. Asif

Wireless Intelligent Networking, Gerry Christensen, Paul G. Florack, and Robert Duncan

Wireless LAN Standards and Applications, Asunción Santamaría and Francisco J. López-Hernández, editors

Wireless Sensor and Ad Hoc Networks Under Diversified Network Scenarios, Subir Kumar Sarkar

Wireless Technician's Handbook, Second Edition, Andrew Miceli

For further information on these and other Artech House titles,including previously considered out-of-print books now available through our In-Print-Forever® (IPF®) program, contact:

Artech House
685 Canton Street
Norwood, MA 02062
Phone: 781-769-9750
Fax: 781-769-6334
e-mail: artech@artechhouse.com

Artech House
16 Sussex Street
London SW1V 4RW UK
Phone: +44 (0)20 7596-8750
Fax: +44 (0)20 7630-0166
e-mail: artech-uk@artechhouse.com

Find us on the World Wide Web at: www.artechhouse.com